ASNT Level II Study Guide

Radiographic Testing Method

by
William Spaulding and George C. Wh[ite]

The American Society for Nondestructive Testing, Inc.

Copyright © 1998 by The American Society
for Nondestructive Testing Inc. (ASNT)

All rights reserved.

Reproduction or transmittal of any part of this book by electronic or mechanical means including photocopying, microfilming, recording, or by any information storage and retrieval system without the expressed written permission from the publisher is prohibited.

Please direct all inquiries to ASNT,
Publications Department, PO Box 28518,
Columbus, OH 43228-0518

ISBN 1-57117-062-6

ASNT is not responsible for the authenticity or accuracy of information herein, and published opinions or statements do not necessarily reflect the opinion of ASNT. Products and/or services that may appear in this book do not carry the endorsement of ASNT. ASNT assumes no responsibility for the safety of persons using the information in this book.

Contents

Introduction 5
   Overview of the Study Guide 5
   Acknowledgments 5
   Recommended References 5
   Resource Materials 5

Chapter 1 – Overview of Radiographic Testing 7
History of Radiographic Testing 7
Advantages and Disadvantages of Radiographic Testing 7
Principles of Radiographic Testing 7
   Types of Penetrating Radiation 8
      X-rays 8
         Electron Source 9
         Electron Target 9
         Electron Acceleration 9
      Gamma Rays 10
         Radiation Energy 10
         Source Activity 10
         Specific Activity 10
         Radiation Intensity 10
         Half Life 11
   Interactions of Radiation with Matter 11
Radiographic Film Exposures 12
   Film Density 12
   Image Quality 12
   Subject Contrast 12
   Effect of Radiation Energy 12
      Effect of Scatter Radiation 12
   Film Contrast 13
   Radiographic Sharpness/Unsharpness 13
      Geometrical Factors Affecting Unsharpness 13
      Nongeometrical Factors Affecting Unsharpness 15
   Scatter Control 15
      Filters 15
      Collimators 16
      Masking 16
   Exposure Reduction and Intensifying Screens 16
   Principles of Shadow Formation 16
      Image Size 16
      Image Shape and Spatial Relationships 17
   Exposure 17
      Exposure Calculations 17
      Exposure Factor 18
         Inverse Square Law 18
         Radiographic Equivalence Factor 18
         Exposure Charts 19
      Thickness, Intensity, Distance, and Time 21
      Variations in Object Thickness 21

Contents

Image Quality Indicators 22
    Identification Markers 23
Film and Film Handling 23
    Film Graininess 23
    Film Selection 23
        Available Forms of Film 24
    Film Handling and Storage 24
    Film Processing 24
        Darkrooms 24
        Manual Processing 24
        Automatic Processing 25
    Exposure Techniques 25
Discontinuity Depth Determinations 27
Interpretation and Evaluation of Radiographs 28
    Visual Acuity and Dark Adaptation 28
    Viewing Conditions 28
    Film Density Measurement 28
    Identifying Discontinuities 29
    Sources of Discontinuities 29
        Inherent Discontinuities 29
        Processing Discontinuities 29
        Service Discontinuities 30
Radiographic Inspection Documents 31
Radiation Safety 31

Chapter 1, Questions 33
Chapter 1, Answers 41

Chapter 2, Procedure Comprehension and Instruction Preparation (PCIP) Examination 43
    Overview of the PCIP Examination Process 43
    Definitions of PCIP Examination Terminology 43
        Specification 43
        Procedure 43
        Instruction 44
            Technique Sheet 44
            Inspection Report 45
        Format of the PCIP Examination 46

Chapter 3, Level II Hands-on Practical Examination 51
    Overview of the Level II Hands-on Practical Examination 51
    General Hands-on Practical Examination 51
    Specific Hands-on Practical Examination 51
    Specific Written Examination 52

Appendix 1, Standard Terminology for Gamma and X-radiography 61

Appendix 2, Qualification and Certification of NDT Personnel 61
    Overview of Personnel Qualification and Certification 61
    Recommended Practice No. SNT-TC-1A 61
    ANSI/ASNT CP-189 63
    ISO 9712 64
    ASNT Central Certification Program (ACCP) 65

Introduction

Overview of the Study Guide

This study guide contains basic information intended to prepare a candidate for Level II radiography examinations administered within the ASNT Central Certification Program (ACCP). This study guide does not present all of the knowledge necessary for certification; the candidate is expected to supplement this guide with the recommended references that follow.

This study guide is divided into three chapters; Chapter 1 and Chapter 2 cover the information found on the Level II General (written) Examination, and Chapter 3 covers the information found on the Level II Hands-on Practical Examination. At the end of Chapter 1, there are questions typical of those that could appear within a portion of the Level II General Examination. These questions contain references for further study, and they are intended to aid the candidate in determining his/her comprehension of the material.

Following each section of Chapter 1 is a "Recommended Reading" box containing references to where additional information on the subjects identified can be found. Listed under "Reference" there is an acronym for a book (*Nondestructive Testing Handbook*, second edition: Volume 3, *Radiography and Radiation Testing* = HB and *Radiographic Testing Classroom Training Book* = CT) followed by a colon and the specific page range where the topic is discussed.

At the end of the study guide, there are several appendices relevant to successful preparation for the ACCP Level II radiography examinations, particularly Appendix 2, which describes pertinent NDT qualification and certification documents allowing for better understanding of the NDT qualification and certification options available to industry.

Acknowledgments

The authors acknowledge the support of the AlliedSignal Aerospace NDT Network and the ASNT Technical Services Department staff. Thanks also to the members of the RT Level II Subcommittee and the RT PQ Methods Committee for their help in the technical review of this study guide.

Recommended References

Nondestructive Testing Handbook, second edition: Volume 3, *Radiography and Radiation Testing* [ASNT order #129]

Radiographic Testing Programmed Instruction Handbook, CT-6-6 (ASNT/General Dynamics) [ASNT order #1612]

Resource Materials

ASM Metals Handbook, 9th edition, Volume 17, *Nondestructive Evaluation and Quality Control, Liquid Penetrant Inspection*, [ASNT order #105]

Recommended Practice No. SNT-TC-1A, 1996 edition [ASNT order #2055]

ANSI/ASNT CP-189-1995: Standard for Qualification and Certification of Nondestructive Testing Personnel [ASNT order #2505]

ASNT Central Certification Program (ACCP) [ASNT order #6001]

Chapter 1
Overview of Radiographic Testing

History of Radiographic Testing

Radiographic testing (RT) unofficially began in 1895 when Wilhelm Roentgen, a German scientist, discovered that an unknown form of radiation emitted from a gas-filled electron tube was capable of penetrating objects that were opaque to light. Roentgen called this radiation "X-rays" but in some countries, X-rays are now called Roentgen rays. At about the same time, Antoine Becquerel, a French scientist, found that radiation from certain uranium compounds had similar properties. This radiation was later determined to be two distinct types called alpha and beta radiation. In 1900, Villard, also a French scientist, found that a third type of radiation emitted from some radioactive materials was similar to X-rays. This radiation was called "gamma rays."

X-rays and gamma rays are essentially the same, differing only in their origin. X-rays are produced artificially by accelerating or decelerating high energy electrons using electronic equipment, while gamma rays are produced by the decay (disintegration of the nuclei) of radioactive isotopes.

Significant use of X-rays and gamma rays for industrial purposes began in the 1920s. Since then, industrial RT has become one of the most commonly used methods of nondestructive testing (NDT). RT is most often used for process control during manufacturing, to detect subsurface discontinuities in end products (i.e., castings, welds, ceramics, and composite composite materials and electronic components) and for quality control inspections of electronic components (determining internal fits, alignments, and/or gaps in assemblies). RT is applied in building and bridge construction, aircraft aviation and aerospace, automotive, and space components manufacturing, and aircraft overhaul, maintenance, and repair. It is also used for inspecting piping and pipelines, refinery vessels, steel pressure vessels, and storage tanks.

Advantages and Disadvantages of Radiographic Testing

RT can be used to detect internal discontinuities in almost any material that is not too thick. X-ray machines capable of penetrating as much as 660 mm (26 in.) of steel, and greater thicknesses of other materials, are available. In addition to discontinuities, RT can disclose internal structures, configurations, fluid levels, and fabrication or assembly errors. In most applications, RT provides an image of the test object that can be kept as a permanent record. Isotopes are often used because of their portability and because they can access hard-to-reach places. They are also used in field testing because they have greater penetrating ability than most portable X-ray machines.

The major limitations of RT are that the opposing sides of the test object must be accessible, precautions to prevent personnel exposure to radiation are required, and configuration of the object must allow for satisfactory formation of shadows of its internal structure.

Principles of Radiographic Testing

RT is based on the detection of differences in the transmission/absorption of penetrating radiation by different parts of the object being radiographed. The differences in transmission may be caused by differences in the thickness or composition of the absorbing material. Sensors that respond to the ionization produced when radiation is absorbed are used to detect the radiation that passes through the object. Special photographic film is the most commonly used detection medium. Just as in photography, the regions of the film where more radiation strikes the film will be blacker after the film is developed. In RT, this produces an image of the part that includes its internal structure and discontinuities. Electronic sensors that react to ionization (i.e., Geiger tubes and scintillation devices) also detect X-rays and gamma rays, and are being used more often in RT applications.

Most discontinuities effectively reduce the thickness of the object, locally, so that more radiation is transmitted at that point. When film is the detector, the transmittal of more radiation results in greater darkening of the film (i.e., the discontinuity image is darker than the remainder of the object). However, some discontinuities, such as tungsten inclusions in welds, may absorb more radiation than the matrix material, which will appear as lighter images on a radiograph. Planar discontinuities (i.e., cracks and lack of penetration) that do not have any appreciable thickness, are different. Because their small thickness causes little change in the amount of radiation that is absorbed, such discontinuities are difficult to detect unless the plane of the discontinuity is nearly parallel to the radiation beam.

Types of Penetrating Radiation

Gamma rays and X-rays are a form of electromagnetic radiation like light, but their energy is much higher and their wavelengths are only about 1/10,000 as large as light. The electromagnetic spectrum shown in Figure 1.1 compares the wavelengths and energies of various forms of radiant energy. The short wavelength and high energy of X-rays and gamma rays enable them to penetrate much more deeply into materials than light can.

The principal characteristics of X-rays and gamma rays are that:

1. the higher their energy, the shorter their wavelength, (wavelength is inversely proportional to energy);
2. they have no mass or electrical charge;
3. they travel at the speed of light;
4. when absorbed or deflected, they ionize matter;
5. the higher their energy, the greater the depth to which they can penetrate in a given material;
6. absorption is increased as the atomic number and density of the absorber increase;
7. they cannot be refracted (as by a lens) or reflected to any useful degree, but they can be diffracted by crystalline structures;
8. living tissue is damaged when it absorbs X-rays or gamma rays.

X-rays

X-rays are produced when rapidly moving electrons are accelerated – either stopped or changed in direction. Usually this is done in a vacuum (X-ray tube) by stopping the electrons with a barrier called a target. This process produces "characteristic X-rays" with energies/wavelengths that depend on the target material, and "Bremsstrahlung" X-rays with energies ranging from near 0 to the maximum energy of the electrons (the voltage at which the X-ray tube was operated). Bremsstrahlung X-rays are also called "white radiation," and it makes up most of the useful radiation in RT. X-ray machines commonly used in RT range in energy from 50,000 electronvolts (50 KeV) to 30,000,000 electronvolts (30 MeV), and in

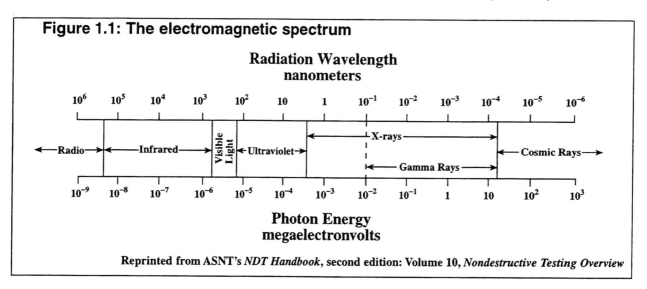

Figure 1.1: The electromagnetic spectrum

Reprinted from ASNT's *NDT Handbook*, second edition: Volume 10, *Nondestructive Testing Overview*

X-ray output from less than 5 to as much as 25,000 rads per minute measured at 1 m (3.3 ft) from the source (roentgens/m-m).

A typical X-ray tube consists of a source of electrons and a target in a vacuum chamber, with the means to apply high voltage across the source-target gap, as shown in Figure 1.2.

Electron Source

When electric current is passed though the wire filament coil in an X-ray tube cathode, the heat generated causes a cloud of electrons to be liberated from the coil. An increase or decrease in the current passing through the filament increases or decreases the number of free electrons, which increases the X-ray output of the tube. The focussing cup helps to keep the electrons bunched together to minimize the size of the focal spot, the area where they strike the target.

Electron Target

The target material must have a high melting point because it becomes very hot when bombarded by electrons from the filament. For the greatest efficiency in producing X-rays, the target should be made of a material with a high atomic number. Tungsten is generally used for the target material because it provides one of the best available combinations of high melting point and high atomic number, although other metals (i.e., copper, iron, cobalt) are used in some X-ray tubes where special applications require particular characteristic radiation.

Electron Acceleration

By applying a negative charge to the cathode and a positive charge to the anode, the negatively charged electrons are repelled by the cathode and attracted to the anode. The higher the voltage difference between the anode and cathode, the higher the velocity of the electrons will be when they strike the target, and the higher the energy of the X-rays that will be generated. Higher energy radiation has greater penetrating power than lower energy radiation. In addition, as the energy of the electrons is increased, the quantity of X-rays generated increases.

In most X-ray machines, the X-ray output is measured indirectly by measuring the tube current (i.e., the flow of electrons from cathode to anode). The values are usually in milliamperes (mA) or microamperes (μA). In some machines, typically those with very high X-ray output (i.e., linear accelerators), the radiation intensity is measured directly in roentgens per minute at 1 m from the target.

Because the production of X-rays is very inefficient, most of the tube current is converted into heat at the target. Consequently, the focal spot size and the cooling of the anode to prevent the target from melting, are major design limitations. In addition, the ability of insulating materials to withstand high voltages

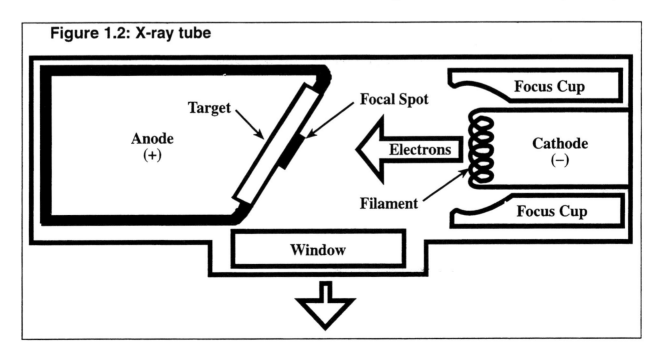

Figure 1.2: X-ray tube

greatly influences tube design. As a result, some machines have a duty cycle rating based on the kilovoltage, tube current, and length of exposure.

X-ray machines are usually rated by their maximum voltage capability in kilovolts or megaelectronvolts. Table 1.1 lists the applications of commonly available X-ray machines and the types of intensifying screens (to be discussed later) that are usually used for each voltage range.

Gamma Rays

Gamma rays are similar to X-rays but they are produced by the decay of naturally occurring or artificially produced radioactive isotopes. Ir-192 and Co-60 are the most commonly used isotopes for RT, and both are artificially produced by neutron bombardment in a nuclear reactor. Until these became available after World War II, naturally occurring radium was used extensively.

The wavelength (energy) of the gamma rays depends on the isotope. Each isotope produces one or more fixed wavelengths, but no Bremsstrahlung radiation.

Radiation Energy

Gamma ray sources are available in many energy ranges from about 10 keV to 12 MeV. The most commonly used energies for Ir-192 are 310 keV, 470 keV, and 600 keV and for Co-60, 1.17 MeV and 1.33 MeV are the most common energies used.

Source Activity

The activity of a gamma ray source depends on the amount of radioactive material present and its rate of decay. The rate of decay is measured in becquerels (curies) and is a useful way of comparing the strength of various sources of the same isotope.

Specific Activity

Specific activity is the activity per unit quantity of the source, expressed as becquerels (curies) per gram. It is useful in RT because a source of a given strength with a high specific activity will be physically smaller than one with a lower specific activity. The smaller source permits a smaller source-to-film distance than a larger source, everything being equal. On the other hand, at the same distance, the smaller source will produce sharper images. (See *Radiographic Sharpness/Unsharpness* on page 13.)

Radiation Intensity

The intensity of radiation from an isotope source (or from a X-ray source) is measured in

Table 1.1: Standard X-ray machine applicable materials and screens

Maximum Voltage	Maximum Application Thickness	Screens
<150 kV	Thin metal sections, electronics, ceramics, plastics	None or lead oxide
150 kV	127 mm (5 in.) aluminum, 25 mm (1 in.) steel Equivalent to 38 mm (1.5 in.) steel	None, lead foil, or lead oxide
250 kV	Equivalent to 51 mm (2 in.) steel Equivalent to 76 mm (3 in.) steel	Lead foil, fluorometallic, or fluorescent
400 kV	Equivalent to 76 mm (3 in.) steel Equivalent to 102 mm (4 in.) steel	Lead foil, fluorometallic, or fluorescent
1 MeV	Equivalent to 127 mm (5 in.) steel Equivalent to 203 mm (8 in.) steel	Lead foil
2 MeV	Equivalent to 203 mm (8 in.) steel	Lead foil or sheets
8-25 MeV	Equivalent to 660 mm (26 in.) steel	Lead foil or sheets

roentgens (R) per unit time at a standard distance from the source. For isotopes, the units are usually roentgens/hour at 1 m (R/hr-m) as compared to X-rays that are usually measured in roentgens/minute at 1 m (R/m-m). The intensity from an X-ray source is also often expressed in terms of the tube current with units such as milliamperes.

Half Life

Because radioactive isotopes decay, the number of active atoms in a source diminishes with time. The time it takes the radioactive material to decay to one half of its initial activity in becquerels (curies) is called the half-life of the isotope. The half-life of Ir-192 is 75 days while Co-60 has a half-life of 5.3 years. This means that 1.9 TBq (50 Ci) of Ir-192 will decay to 1 TBq (25 Ci) in 75 days, to 0.5 TBq (12.5 Ci) in the next 75 days, and so on.

Interactions of Radiation with Matter

X-rays and gamma rays have no mass or weight – they are bundles of energy called photons traveling at the speed of light. They can be absorbed or deflected by matter in a number of ways, usually by causing atoms of the matter to become ionized (electrically charged). Electrons and/or lower energy photons are emitted from the atom in a different direction from that of the incident photon. These electrons or photons may, in turn, cause the ionization of other atoms in the absorber.

Some of the major processes that account for absorption/deflection are the photoelectric effect, Compton scattering, and pair production. The photoelectric effect is most important for photon energies up to about 0.3 MeV. From 0.3 to 1.3 MeV, Compton scattering prevails, and at higher energies, pair production is dominant.

Recommended Reading

Subject	Reference*
history of radiographic testing	HB: 115
advantages and disadvantages of radiographic testing	HB: 30, 50; CT: 1-4
X-rays	HB: 71-72, 93, 96; CT: 2-7 to 2-18, 3-7
gamma rays	HB: 71, 73, 461; CT: 2-18 to 2-22
radiation's interaction with matter	HB: 65, 76-77, 265; CT: 2-13, 2-16

*See *Introduction* for explanation of references.

Radiographic Film Exposures

A radiograph is the shadow picture produced by X- or gamma radiation that has passed through an object and been partly absorbed by film. (**Note:** Most of the radiation reaching the film passes through the object and is absorbed by other objects.) The radiation that is absorbed in the film sensitizes the silver halides in the film emulsion in such a way that a chemical process called development can convert them to silver particles. In the film areas exposed to higher levels of radiation, more silver particles are produced by development, making the film darker, while film areas exposed to less radiation, due to a thicker or more absorptive object material, are lighter after processing, as shown schematically in Figure 1.3.

Film Density

The degree of film darkening is called film density, which is measured by the amount of visible light that can penetrate the film. Density is defined as the logarithm of the amount of light incident on one side of the film divided by the amount of light transmitted through the film. Mathematically, density is presented as:

$$D = \log (I_o/I_t) \quad \text{(Eq. 1)}$$

where:
D = density
I_o = light intensity incident on the film
I_t = light intensity transmitted

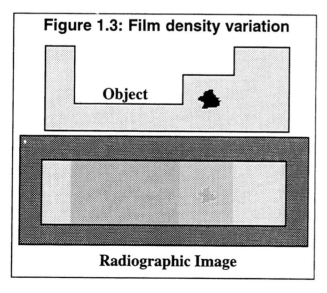

Figure 1.3: Film density variation

Image Quality

The usefulness of any radiograph depends on the quality of the image (sensitivity). Sensitivity is defined as the smallest detail of the object that can be seen on the radiograph. It is a function of the contrast and the sharpness (definition) of the radiographic image.

Radiographic contrast is the difference between the film densities of two areas of a radiograph. This overall contrast depends on the contrast provided by the object being radiographed and the contrast provided by the film. These are usually referred to as subject contrast and film contrast, respectively.

The sharpness is usually judged from the image of known features such as edges, steps, or holes in the object. Sharpness is a function of geometric factors such as source size, source-to-film distance, object-to-film distance, and screen-to-film contact, as well as type of film and screens, and the radiation energy used. Usually, unsharpness is called geometric unsharpness, because that is the component of sharpness that can be calculated.

Subject Contrast

Subject contrast is governed by the ratio of the intensity of transmitted radiation through various parts of the object and by the amount of scatter radiation reaching the film. The relative amounts of radiation transmitted through various regions of a specimen depend on the thicknesses of those regions and on the radiation energy being used. Large differences in thickness produce high subject contrast, and vice versa.

Effect of Radiation Energy

Low energy radiation also produces high subject contrast, because it is more easily absorbed than high energy radiation. Therefore, only a small change in thickness radiographed when using low energy radiation is necessary to achieve reasonable contrast. On the other hand, a large thickness difference is necessary to achieve reasonable contrast when using high energy radiation.

Effect of Scatter Radiation

Scatter radiation (scatter) is the secondary radiation that results from radiation interactions with matter, as previously described. Although the primary radiation emitted by the source travels in straight lines, when it interacts with matter, the resulting secondary radiation travels in all directions. The scatter has lower energy than the primary radiation, so each photon and/or electron of scatter is easier for the film to absorb and has greater effect on the density of the radiographic image. Because scatter radiation travels in all directions, it does not produce a useful image of the object being radiographed – it merely fogs or darkens the film overall, reducing the contrast. The effects of scatter can be controlled by use of filters and masking, which will be discussed in more detail later.

Film Contrast

Each type of radiographic film has a characteristic relationship between the amount of exposure and the density that is produced by that exposure. The relationship is usually expressed as a graph or characteristic curve in which the density is plotted against the logarithm of the relative exposure, as shown in Figure 1.4. Relative exposure is used because there are no other measurement units that apply to all possible exposure conditions. The log of the relative exposure is used to compress an otherwise long scale. A log scale has the added value that the same distance will separate the logs of any two exposures having the same ratio on a log scale, regardless of the actual exposure values. This feature is useful in exposure calculations.

The slope of a film's characteristic curve is a measure of its contrast, while the curve's position left or right within the graph is a measure of film speed. The contrast is greatest where the greatest density difference is produced by a given difference in exposure, that is where the slope of the curve is greatest.

Given the same degree of development, a film with a curve that lies to the left of another film is the faster film, because the left-most curve indicates that less exposure is required to produce a given density. The shape of the characteristic curve of a given film is not sensitive to radiation energy but it is affected by the degree of film development – time, temperature, and composition of the developer. Within limits, an increase in degree of development increases the contrast exhibited by the radiograph.

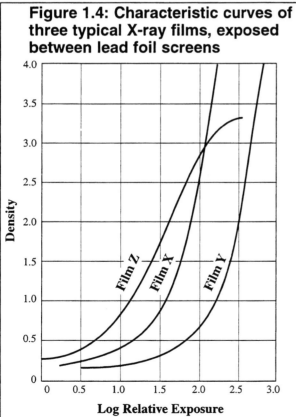

Figure 1.4: Characteristic curves of three typical X-ray films, exposed between lead foil screens

Reprinted from ASNT's *NDT Handbook*, second edition: Volume 10, *Nondestructive Testing Overview*

Radiographic Sharpness/Unsharpness

Source-to-film distance is a major factor in the production of quality, cost-effective radiographs. If the source-to-film distance is too small, critical discontinuities such as cracks may not be visible at all because of image unsharpness. If the source-to-film distance is larger than necessary for good image sharpness, exposure time will be needlessly long, increasing the costs of the process. Unsharpness can be caused by geometrical and nongeometrical factors.

Geometrical Factors Affecting Unsharpness

Because radiographs are shadow images, radiographers must understand the geometric factors that influence shadow formation. Shadows consist of an umbra and a penumbra. The umbra is the central, darkest part of the shadow; the penumbra, usually called the geometric unsharpness (U_g), is the lighter, fuzzy shadow surrounding the umbra. Under most practical radiographic conditions, only the umbra is visible so it is very important to maximize its size. For objects smaller than the radiation source, the umbra can only be maximized by minimizing the geometric unsharpness. Because many discontinuities are smaller than the source, particularly the width of cracks, this process frequently used. For practical reasons, many specifications permit more than the minimum attainable unsharpness. Generally acceptable values range from 0.1-0.8 mm (0.005-0.03 in.).

Figure 1.5 illustrates two cases of geometric unsharpness while Equations 2 and 3 provide two forms of the equation used to calculate unsharpness. From the figure or the equations, it is evident that the following steps are necessary to minimize the geometrical unsharpness of a radiograph:

1. the focal spot of the X-ray tube or the physical size of the isotope source (F) should be as small as possible within the limits imposed by the need for sufficient radiation output;
2. the distance from the source to the source side of the test object (source-to-object distance) should be as large as possible within the limits imposed by the need for economical exposure time;
3. the distance from the source-side of the object to the film (object-to-film distance) should be as small as possible

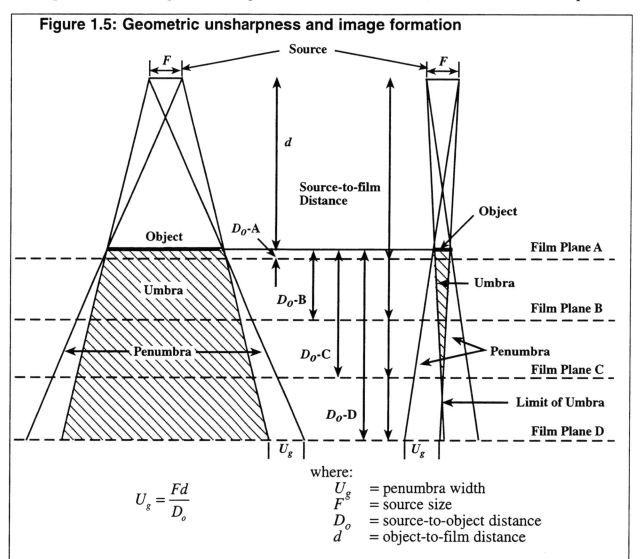

Figure 1.5: Geometric unsharpness and image formation

$$U_g = \frac{Fd}{D_o}$$

where:
- U_g = penumbra width
- F = source size
- D_o = source-to-object distance
- d = object-to-film distance

NOTE: If the object is smaller than the source, umbra (image) will always be smaller than the object, and if object-to-film distance exceeds a critical value, no image may appear on the film. This can occur with most cracks because crack thickness is usually smaller than any source size. Objects larger than the source are always magnified.

(i.e., the film holder should be in direct contact with the object whenever possible.)

$$\frac{U_g}{F} = \frac{d}{D_o} \quad \text{(Eq. 2)}$$

$$U_g = F\frac{d}{D_o} \quad \text{(Eq. 3)}$$

where:
- U_g = geometric unsharpness
- F = effective focal spot size
- D_o = source-to-object distance
- d = object-to-film distance

Note: When the film is in contact with the object, the object-to-film distance is equal to the thickness of the object. If the film is not in contact with the object, it is mandatory to use the actual object-to-film distance, not the thickness of the object. All dimensions must be in the same units, usually millimeters (inches).

The effective source size to be used in the equations is the largest dimension of the focal spot of an X-ray tube or the radioactive portion of an isotope source, as seen from the film; it may be the diagonal of a rectangular or cylindrical source. The manufacturer of the source can provide the nominal source size and shape, but it is the responsibility of the radiographer to determine the projected source size (i.e., the size as seen from the film). This is done by calculating the shape and dimensions of the source.

The sizes and shapes of focal spots are usually determined experimentally by the manufacturer for a given design of machine. For energies below 500 keV, the pinhole technique is usually used, while for higher energies, one of the laminated-collimator techniques is used.

Commonly used X-ray machines have nominal focal spot sizes of 1 mm (0.04 in.) to as much as 8 mm (0.3 in.), in the shape of rectangles, ellipses, or circles. Gamma ray sources vary widely in size and shape, depending on the form of the radioactive material and the source strength.

Some special purpose X-ray machines have a focal spot size that is less than 0.1 mm (0.004 in.). These machines can be used with relatively large object-to-film distances to provide considerable magnification of the image without significant loss of image sharpness. Because the small focal spot of X-ray machines severely limits the radiation output, they are typically used only for the RT of objects with low radiation absorption.

Nongeometrical Factors Affecting Unsharpness

Unsharpness is also affected by the radiation energy, the type of film, or other detector used, the type and position of screens and filters used, and the contact of the screens with the film. Increasing the radiation energy or increasing the speed of the film being used increases the unsharpness. Fluorescent screens increase unsharpness, while metallic screens vary in their effects depending on their composition, thickness, and position relative to the film. All screens and filters that are within the film holder MUST be in intimate contact with the film or they will greatly increase unsharpness.

Scatter Control

Because scatter radiation reduces contrast, it must be prevented from reaching the film, insofar as practical. The most common ways to reduce scatter are with filters, collimators, and masking.

Filters

The most common technique used to reduce scatter is by absorbing the scatter with filters near the film. (**Note:** filters are often called "screens" because many common screens also reduce scatter.) Because scatter radiation is less penetrating than the primary radiation from the source, useful filtering can be accomplished with relatively thin sheets of an absorber – lead is the most common material. Many commercial film holders are provided with a 0.3 mm (0.01 in.) lead "back filter" to absorb scatter from material behind the holder, such as the floor or table on which the object and film are resting. For high energy RT, thicker back filters are usually necessary. For many applications, front filters are often useful, but the details are usually specific to the application.

Collimators

Reducing the generation of scatter is also an important control. This may be done by filtering or collimating the primary radiation beam or by masking the object. Filters and collimators reduce scatter in the entire radiation area by removing much of the less penetrating (softer) primary radiation while leaving greater amounts of the more penetrating (harder) radiation needed to produce the radiographic image. Collimating limits the primary beam width by removing the parts of the primary beam that would otherwise pass around the object and generate scatter.

Masking

Masking is performed by fitting absorptive material closely around the test object. Masking is similar to collimating in that it limits the primary beam to the object. The absorptive masking material is often made of lead sheets formed to the object, but lead, copper, or steel shot, and even some liquids may be useful in specific cases.

Exposure Reduction and Intensifying Screens

The radiographic exposure time required to produce a desired image density can be shortened in many cases by placing intensifying screens in close contact with the film. Intensifying screens are constructed of materials that, when struck by the primary radiation, produce secondary radiation that blackens the film more effectively than the primary radiation. Fluorescent screens produce light, while metallic intensifying screens produce electrons and secondary X-ray photons. Fluorescent screens are seldom used with film because they reduce the definition of the image. An exception is when very long exposure times are needed.

For RT with 150 kV radiation or higher, lead is the most common material for metallic screens, although other metals may be useful in some cases. Lead screens 0.03-0.3 mm (0.001-0.01 in.) thick are used up to 1-2 MeV. For higher energies, lead screens as much as 6 mm (0.25 in.) thick have been found to be useful. Metallic screens may be used on one or both sides of the film, and when two or more films are used in the same holder, thin screens are sometimes used between the films. In all cases, the screen surface must be kept very clean and free from foreign materials (i.e., grease and dust) because they can cause artifacts in the radiograph. Below 150 kV, a thin layer of lead oxide on a supporting material is often useful for intensification.

Lead and lead oxide screens can shorten exposure times by as much as 2-2.5 times. Using lead and lead oxide screens with other metallic screens, also provides useful filtration of scatter. Screens reduce exposure time and increase contrast, and should be used in most RT applications.

Principles of Shadow Formation

In order to provide useful images of the object to the interpreter so an accurate interpretation of radiographs can be made, the radiographer must consider the principles of shadow formation in making the radiograph. Because a radiograph is a shadow image of an object placed between the radiation source and the recording medium, the shape, size, and spatial relations of the parts on the image are influenced by the relative positions of the film (or other detector), the object, and the source.

Image Size

If the source of radiation is larger than the object, the image of the object will be smaller than the object except when the object is in contact with the film. While most objects being radiographed are not smaller than the source, the principal objective of much RT is to detect discontinuities. Since discontinuities may well be smaller than the source, and the thickness of cracks and lack of penetration is almost always smaller than the source, this rule of shadow formation is very important in searching for discontinuities.

If the source is smaller than the object, the image of the object will be larger than the object except when the object is in contact with the film. This is the case with penetrameter images, and because penetrameters have a known size and are placed on the source side of the object, the size of the penetrameter image provides a useful means for estimating the source-to-object distance when the source size is known. Of course, when the penetrameter is placed on the film, this technique is useless.

The degree of enlargement may be calculated mathematically using Equation 4

$$\frac{S_o}{S_i} = \frac{D_o}{D_i} \qquad \text{(Eq. 4)}$$

where:
- S_o = object size
- S_i = image size
- D_o = source-to-object distance
- D_i = source-to-film distance

Image Shape and Spatial Relationships

Image distortion occurs when the radiographic image is not the same shape as the object or discontinuity that produced it. Image distortion occurs when the plane of the object and the plane of the film are not parallel. Usually, the preferred practice is to keep the film plane as parallel as possible to the plane of the object that is of maximum interest, even though this may distort the image of other portions of the object. **Note:** this does not mean that suspected cracks or lack of penetration should be oriented parallel to the film; they should be parallel to the radiation beam for reasons discussed previously in the *Radiographic Sharpness/Unsharpness* section (page 13).

Distortion of spatial relationships between parts of the object may also occur, as shown in Figure 1.6(a). For this reason, the preferred practice for most RT problems is to keep the center of the radiation beam perpendicular to film as shown in Figure 1.6(b).

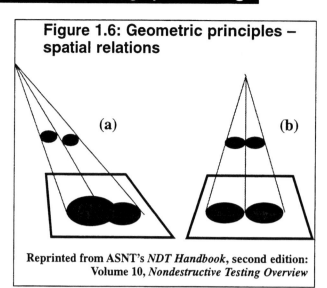

Figure 1.6: Geometric principles – spatial relations

Reprinted from ASNT's *NDT Handbook*, second edition: Volume 10, *Nondestructive Testing Overview*

Exposure

Radiographic exposure is defined as the intensity of the radiation multiplied by the time that the film is exposed to the radiation. For a given radiation energy and source-to-film distance, the exposure may be stated as

$$E = MT \qquad \text{(Eq. 5)}$$

where:
- E = exposure
- M = radiation intensity [tube current or roentgens/m, or becquerels (curies)]
- T = time

Because the amount of radiation that is reaching the film through the object is unknown, the intensity used in the equation is the intensity of the X-ray or isotope source, as measured in becquerels (curies), rads/minute, or tube current units.

For instance, an exposure at 5 mA for 10 minutes would be equal to an exposure of 10 mA for 5 minutes. The units of exposure are determined by the units used for radiation intensity and time. In this example, exposure would have the units of milliampere-minutes. Values such as milliampere-seconds, becquerel (curie)-minutes, and rads are also common.

While the output of X-ray machines can be selected by the operator within the limits of the machine, the radiographer must consider the half-life of isotope sources in determining their output at any given time. A satisfactory exposure of 3.2 TBq-min (100 Ci-min) that required 1 minute when made 75 days ago with a 3.7 TBq (100 Ci) Ir-192 source, will now require 2 minutes at the same source-to-film distance because the source has decayed to 1.9 TBq (50 Ci).

Exposure Calculations

The density of a radiograph depends on the amount of radiation absorbed by the film emulsion and how it was developed. The amount of radiation absorbed depends on the amount and energy of the radiation source (primary radiation) that passes through the object, the amount of scatter reaching the film, and the action of any intensifying screens that were used.

For any given radiation energy, the controllable variables that govern exposure are the source output in rads/minute, the time that the film is exposed, and the source-to-film distance. Because the output of X-ray machines is proportional to the tube current, milliamperes or microamperes may be used when output in rads/minute is not available. For gamma ray sources, output is measured in becquerels (curies).

Exposure Factor

To make exposure values more general, the source-to-film distance may be factored into the exposure equation to provide what is called an "exposure factor." As shown in Equation 6 (X-ray) and Equation 7 (gamma ray), the exposure factor is the E value divided by the square of the source-to-film distance. For example, an exposure of 400 mAm at 508 mm (20 in.) source-to-film distance has an exposure factor of 1 mAm/in.^{2}, as does an exposure of 100 mAm at 254 mm (10 in.) source-to-film distance.

$$EF_x = \frac{M(t)}{D_i^2} \qquad \text{(Eq. 6)}$$

$$EF_r = \frac{S(t)}{D_i^2} \qquad \text{(Eq. 7)}$$

where:
- EF = exposure factor
- D_i = source-to-film distance
- M = X-ray tube current
- t = time
- S = gamma ray source strength

Inverse Square Law

When no absorber is present (e.g., in air), the radiation intensity from any radiation source decreases as the square of the distance from the source increases. In other words, the intensity is inversely proportional to the square of the distance from the source. This occurs because the radiation diverges as it travels away from the source, so that the same amount of radiation covers a larger area. Thus, the radiation is less intense farther from the source. Figure 1.7 illustrates this effect, which is known as the inverse square law.

Mathematically, this law is expressed as

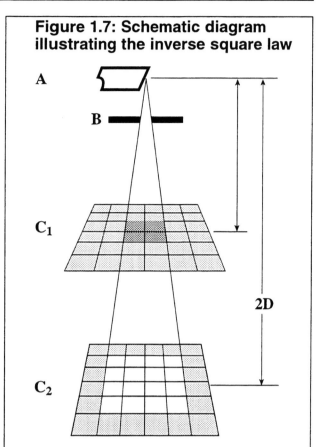

Figure 1.7: Schematic diagram illustrating the inverse square law

Reprinted from ASNT's *NDT Handbook*, second edition: Volume 10, *Nondestructive Testing Overview*

$$\frac{I_1}{I_2} = \frac{(D_2)^2}{(D_1)^2} \quad \text{or} \quad I_2 = I_1 \frac{(D_1)^2}{(D_2)^2}$$

where:
- I = intensity
- D = distance

The inverse square law is very important in RT because different source-to-film distances are often used for different radiographs. Source-to-film distance changes may be needed to satisfy image unsharpness requirements, allow for coverage of the object in one exposure, adjust radiation intensity so as to adjust exposure time, and similar reasons.

Radiographic Equivalence Factor

While many RT operations are performed on one type of material, others are sometimes used to radiograph unusual materials. When unusual materials are encountered, it is useful to have some means of determining exposures for the new material based on exposure data for the well-known material. This can be done with a chart or table of radiographic equivalence factors[b] that is similar to Table 1.2.

To use the table, choose the radiation energy of interest and the corresponding material of interest. Multiply the resulting equivalence factor by the thickness of the material to be radiographed. This results in a thickness value of either aluminum or steel, depending on the radiation energy, that has approximately the same absorption as the thickness of new material. For example, at 220 kV, 13 mm (0.5 in.) of copper is equivalent in absorption to 13 mm (0.5 in.) × 1.4 = 18 mm (0.7 in.) of steel. To radiograph 13 mm (0.5 in.) of copper at 220 kV, use the same exposure that was used for 18 mm (0.7 in.) of steel.

Exposure Charts

Exposure charts are a means of simplifying the selection of the proper values of the variables needed to produce acceptable radiographs. A common type of exposure chart is shown in Figure 1.8[1]. As in all such charts, certain variables of RT have been fixed or predetermined. In Figure 1.8, the fixed variables are film type, source-to-film distance, screen type and thickness, and desired film density, while kilovoltage, material thickness and exposure are controllable variables.

The chart simplifies the relationships between material thickness, kilovoltage, and exposure by fixing the other variables. This makes it easier for the operator to select exposure values. For example, in the RT of a 25 mm (1 in.) thick steel part with this X-ray machine, 220 kV, 200 kV, 180 kV or 160 kV might be chosen. The chart shows that exposures would range from about 5.3 mAm at 220 kV to 70 mAm at 160 kV. If the X-ray machine were operated at 5 mA, the required exposure time at 160 kV would be 14 minutes, while at 220 kV, the time would be 1.3 minutes.

Charts in which other variables are fixed are also useful. For example, if the source-to-film distance, film, screens, density, film processing, and exposure are fixed, a chart can be prepared

Table 1.2: Approximate Radiographic equivalence factors[a]

Material	X-rays (kilovolts)								Gamma Rays			
	50	100	150	220	400	1000	2000	4 to 25	Ir-192	Cs-137	Co-60	Radium
Magnesium	0.6	0.6	0.05	0.08								
Aluminum	1.0	1.0	0.12	0.18					0.35	0.35	0.35	0.40
2024 (aluminum) alloy	2.2	1.6	0.16	0.22					0.35	0.35	0.35	
Titanium			0.45	0.35								
Steel		12.0	1.0	1.0	1.0	1.0	1.0	1000	1.0	1.0	1.0	1.0
18-8 (steel) alloy		12.0	1.0	1.0	1.0	1.0	1.0	1000	1.0	1.0	1.0	1.0
Copper		18.0	1.6	1.4	1.4			1300	1.1	1.1	1.1	1.1
Zinc			1.4	1.3	1.3			1200	1.1	1.0	1.0	1.0
Brass[b]			1.4	1.3	1.3	1.2	1.2	1200	1.1	1.1	1.1	1.1
Inconel X alloy-coated		16.0	1.4	1.3	1.3	1.3	1.3	1300	1.3	1.3	1.3	1.3
Zirconium			2.3	2.0	1.0							
Lead			14.0	12.0		5.0	2.5	3000	4.0	3.2	2.3	2.0
Uranium				25.0				3900	12.6	5.6	3.4	

[a] Aluminum is the standard metal at 50 kV and 100 kV and steel at the higher voltages and gamma rays. The thickness of another metal is multiplied by the corresponding factor to obtain the approximate equivalent thickness of the standard metal. The exposure applying to this thickness of the standard metal is used. EXAMPLE: To radiograph 12.7 mm (0.5 in.) of copper at 220 kV, multiply 12.7 mm (0.5 in.) by the factor 1.4, obtaining an equivalent thickness of 17.8 mm (0.7 in.) of steel.

[b] Tin or lead alloyed in brass will increase these factors.

Reprinted from ASNT's *NDT Handbook*, second edition: Volume 10, *Nondestructive Testing Overview*

---

[1] Although not stated, this chart applies only to one specific X-ray machine and the film processing time, temperature, and chemicals are also fixed.

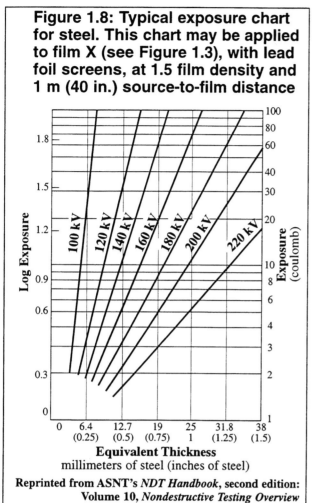

Figure 1.8: Typical exposure chart for steel. This chart may be applied to film X (see Figure 1.3), with lead foil screens, at 1.5 film density and 1 m (40 in.) source-to-film distance

Reprinted from ASNT's *NDT Handbook*, second edition: Volume 10, *Nondestructive Testing Overview*

Figure 1.9: Typical gamma ray exposure chart for Ir-192, based upon the use of film X (see Figure 1.3)

Reprinted from ASNT's *NDT Handbook*, second edition: Volume 10, *Nondestructive Testing Overview*

that shows the relationship between kilovoltage and thickness for a given material.

Exposure charts for isotopes typically plot the exposure factor (rather than the exposure) against material thickness for various resulting film densities as shown in Figure 1.9. This type of chart reflects the need to allow for source-to-film distance variance to accommodate different thicknesses, because the operator cannot control the energy or the output of an isotope.

Deviations from some fixed variables of an exposure chart can be compensated for mathematically as follows:
1. source-to-film distance – use the inverse square law;
2. film type – use the characteristic curves of the films;
3. desired film density – use the characteristic curve of the film.
4. film processing – if characteristic curves for other temperatures, chemicals or development times are available, use them; otherwise, see item 2, below.

Changes to the following variables of the exposure chart cannot be accurately predicted:
1. X-ray machine – all X-ray machines are different; two X-ray machines operating at the same nominal kilovoltage and tube current may produce significantly different energies and intensities of radiation;
2. film processing – a change in chemicals, temperature, or development time will change the resulting film density and contrast;
3. type or thickness of screens or filters – any change in the energy spectrum of the radiation reaching the film, such as those produced by screens and filters, may change the density and/or contrast of the resulting radiographs.

Thickness, Intensity, Distance, and Time

The relationships between object thickness, source intensity, source-to-film distance, and exposure time are mathematical and require that calculations be made or that the radiographer interpret charts. The calculations for changes in source-to-film distance, intensity, or time are simple arithmetic functions as demonstrated in Figure 1.10.

Variations in Object Thickness

An acceptable exposure of an object with varying thicknesses requires intelligent use of the RT variables. The use of filters or higher radiation energy will reduce contrast, but sensitivity may become unsatisfactory. Reductions in energy or removal of filters will increase contrast. The use of slower films increases contrast and sharpness, while faster films reduce contrast and sharpness.

For high subject contrast situations, two or more films of the same or different speeds may be exposed simultaneously in the same film holder. The parts of the image showing suitable density on any one film are interpreted using a single film while two or more films are superimposed in order to view the lower density regions of the image.

Two films, usually of different speeds, that are loaded and exposed together in a single

---

Figure 1.10: Sample calculations

Sample 1 is based on the following: initial exposure is 2 min., 5 mA, 20 in. source-to-film distance. It is desired to change the source-to-film distance to 36 in.
General Rule: The current, in milliamperes, (M) is directly proportional to the square of the source-to-film distance (D).

$$\frac{M_1}{M_2} = \frac{(D_1)^2}{(D_2)^2} \quad \frac{5}{x} = \frac{20^2}{36^2} \rightarrow \frac{5}{x} = \frac{400}{1296} \rightarrow \frac{1}{5}\left(\frac{5}{x}\right) = \frac{1}{5}\left(\frac{400}{1296}\right) \rightarrow x \frac{6480}{400} = 16.2 \text{ mA}$$

Sample 2 is based on the following: initial exposure is 2 min., 5 mA, 20 in. source-to-film distance. It is desired to change the source-to-film distance to 36 in.
General Rule: The exposure time (T) is directly proportional to the square of the source-to-film distance (D).

$$\frac{T_1}{T_2} = \frac{(D_1)^2}{(D_2)^2} \quad \frac{2}{x} = \frac{20^2}{36^2} \rightarrow \frac{2}{x} = \frac{400}{1296} \rightarrow \frac{1}{2}\left(\frac{2}{x}\right) = \frac{1}{2}\left(\frac{400}{1296}\right) \rightarrow x \frac{2592}{400} = 6.48 \text{ min}$$

Sample 3
General Rule: The current, in milliamperes, (M) required is inversely proportional to time (T). Using the results from Sample 1 and Sample 2, calculate for 10 mA.

$$\frac{M_1}{M_2} = \frac{T_2}{T_1} \quad \frac{16.2}{10} = \frac{x}{6.48} \rightarrow \frac{x}{6.48} = \frac{16.2}{10} \rightarrow \frac{6.48}{1}\left(\frac{x}{6.48}\right) = \frac{6.48}{1}\left(\frac{16.2}{10}\right) \rightarrow x \frac{105}{10} = 10.5 \text{ min}$$

or

$$M_1 \times T_1 = M_2 \times T_2 \rightarrow 16.2 \times 6.48 = 10x \rightarrow 104.98 = 10x \rightarrow \frac{104.98}{10} = \frac{10x}{10} \rightarrow 10.498 = x = 10.5 \text{ min}$$

Sample 4 is based on the following: initial exposure is 3.4 min., 75 Ci Ir-192 source, 18 in. source-to-film distance. Calculate the time required for 30 Ci.
General Rule: Time (T) is inversely proportional to source strength (S).

$$S_1 \times T_1 = S_2 \times T_2 \rightarrow 75 \times 3.5 = 30(T_2) \rightarrow 262.5 = 30(T_2) \rightarrow \frac{262.5}{30} = T_2 = 8.75 \text{ min}$$

holder are sometimes viewed separately. The advantages of this technique are evident in situations where the material thickness or absorptivity cannot be precisely determined or where the object contains large differences in thickness.

Double or triple loading refers to the number of radiographic films placed in a single film holder. Different film speeds are used to obtain acceptable film densities over a wide range of cross sectional thicknesses. Each film effectively images a separate area of interest that, when combined, provides total coverage of the object and enhances latitude.

It is especially important in the RT of multi-thickness parts to understand thoroughly how to use the characteristic curves of films and the exposure charts for the available radiation sources. Proper use of these aids can greatly increase efficiency and reduce the costs and time for RT.

Recommended Reading

Subject	Reference*
radiographic film exposures	HB: 188; CT: 2-16
radiographic film density	HB: 179, 384; CT: 4-6
subject contrast	HB: 227, 322; CT: 4-3
film contrast	HB: 224-226
radiographic sharpness/unsharpness	HB: 264; ASTM E 1165
filters and masking	HB: 209, 663; CT: 6-8, 6-11
shadow formation	HB: 192
intensifying screens	HB: 276; CT: 6-9, 6-30
inverse square law	HB: 197; CT: 2-11, 6-24
exposure charts	HB: 222
characteristics curves/calculations	HB: 221-226

*See *Introduction* for explanation of references.

Image Quality Indicators

Image quality indicators (IQIs) provide assurance that satisfactory radiographic image quality has been obtained. The most commonly used type of IQI is the penetrameter. It is a small test piece of standard design, made of material that is radiographically similar to the object, that is radiographed together with the object. It is placed on the source-side of the object whenever possible, so that its image represents the largest object-to-film distance, and thus the largest unsharpness, displayed by that radiograph. It is a good practice to provide the image of at least one penetrameter on each radiograph and more often than not, it is required.

There are many standard designs of penetrameters. The most commonly used designs are small shims (plaques) containing holes and sets of small diameter wires. The dimensions of penetrameter features are some small percentage of the thickness of the object, and image quality is judged by the smallest visible penetrameter feature, such as hole size or wire diameter. **Important:** during typical usage, penetrameters do not provide positive measurement of image quality. The penetrameter image on a radiograph indicates only that the image quality is not poorer than some minimum requirement.

There are two plaque-type penetrameters commonly used in the United States,

ASTM/ASME penetrameters conforming to ASTM E 1025 and MIL-STD-453 penetrameters. Wire penetrameters conforming to ASTM E 747 are gaining in popularity in the United States. These are similar (but not identical) to the DIN or ISO penetrameters widely used in Europe. The image quality or sensitivity values obtained from the various types of penetrameters are not identical, but they are mathematically related. The relationship for ASTM plaque and wire types is charted in ASTM E 747.

When specifications require particular types or sizes of penetrameters that are not readily available, it is useful to be able to determine the characteristics of equivalent penetrameters. For plaque penetrameters, ASTM E 1025 provides an equation and a nomogram for determining equivalent penetrameter sensitivity.

The penetrameter image is not intended to be used to judge the size or acceptability of discontinuities.

Identification Markers

Radiographs must be marked in such a way that each one can be identified with the object that it represents. For objects requiring more than one radiograph, each one must be identified with the part of the object that it represents, so the film can be matched to the corresponding region of the object. Lead letters and numbers placed on the object are usually used for this purpose because their high radiographic absorption allows them to be imaged on the radiograph. The exact locations of the lead markers may be permanently marked on the object or their locations may be keyed to a map of the object and retained as a permanent record.

Identification and location markers are obviously essential in order to correlate the radiographic images of any discontinuities with their location in the object. Specific requirements for marking vary considerably from customer to customer. Typical requirements are available in standards such as ASTM E 94, ASTM E 1030, various ASME codes, and other specifications.

Film and Film Handling

Industrial radiographic film consists of a thin sheet of transparent plastic called the "base" that is coated, usually on both sides, with photosensitive material called the "emulsion." The emulsion is a solid, gelatinous material approximately 0.03 mm (0.001 in.) thick containing microscopic particles of silver halide. When the silver halide absorbs electromagnetic radiation, including visible light, it is modified so that the chemicals present in photographic developer can change the silver halide to metallic silver. The developer does not change the silver halide that did not absorb radiation. After developing, the remaining halide is removed by photographic fixer, leaving just the metallic silver. Areas of the emulsion that contain little silver are relatively transparent to light, while those where there is much silver are less transparent, or denser.

Film Graininess

Microscopic grains of silver form the radiographic image. However, for various reasons, these particles tend to clump together in relatively large masses that are sometimes visible to the naked eye as "graininess." All films exhibit graininess to some degree. Slow speed, fine grain films exhibit lower levels of graininess and higher definition. Graininess is reduced when the radiation energy that produced the image is low; increasing the radiation energy increases the graininess. Graininess is also be affected by the film development process.

Film Selection

Choosing the right film for a particular application is the radiographer's responsibility. The composition, size, and thickness of the object, energy, and output of the radiation source, the criticality of the inspection, and the required level of sensitivity must be considered when selecting the type of film. The time and cost saving advantages of higher speed films must be weighed against their poorer contrast and sharpness as compared to slower films. The possibility of using relatively slow film but decreasing the exposure time by using intensifying screens must be considered. However, fluorescent intensifying screens should only be used when the highest possible photographic speed is required and the lower sensitivity can be tolerated.

Available Forms of Film

Industrial radiographic film is typically available in individual sheets with or without interleaves (separating paper) in a variety of sizes ranging from 127 × 178 to 356 × 432 mm (5 × 7 to 14 × 17 in.). When smaller pieces are needed for insertion into confined spaces, such film is easily cut in the darkroom; of course it must be inserted into a light-tight envelope, usually made of black plastic sheet, and sealed with black tape before leaving the darkroom.

Pre-packaged film is available in sealed, light-tight envelopes, with or without lead oxide screens, and it is ready for exposure without removing it from the envelope. The advantages of pre-packaged films are the elimination of the time for loading film holders, and the convenience of using it in situations where a darkroom is not readily available (i.e., field RT). However, compared to standard sheet film, pre-packaged film is expensive, so it should be used only where the advantages justify its cost.

Film is also available in long rolls. Roll film is advantageous for inspecting large circumferential welds or other cylindrical objects. The film is wrapped around the outside of the cylinder while the radiation source is centered inside. Unless the object is extremely large, only one exposure is needed with this technique. The advantages of roll film for such work include reduction of the required setup time and reduction of the number of identification and location markers required.

Film Handling and Storage

Film must be handled carefully to avoid damaging the emulsion layers. Pressure marks, creases, finger prints, scratches, static marks, humidity, heat, moist/contaminated hands, and splashes or spills of processing chemicals can produce artifacts that may render a radiograph unacceptable.

To avoid problems, always wear cotton gloves when handling dry film, handle it only by the edges, slide it slowly (not rapidly) out of its box or film holder, keep processing chemicals away from the loading bench, and promptly wipe up any spills or splashes. Store films in a cool, preferably air-conditioned location, away from penetrating radiation, and store only the amount of film that can be used by the film expiration date.

Film Processing

After exposure, film must be processed to develop the image and fix it so that the image will not deteriorate as it ages. Both developing and fixing are chemical processes that must take place in a darkroom or other location where there is little light or other radiation. Developer chemicals are alkaline organic compounds that convert exposed silver halide into silver, while fixers are acidic inorganic compounds that convert the remaining silver halide into compounds that can be dissolved in water so that they can be removed from the film. Both developers and fixers also have other functions such as hardening the emulsion so that is not easily damaged during processing and subsequent handling.

Darkrooms

Darkrooms vary in size and layout, but all must be lightproof, radiation free, equipped with safelights, and have a convenient, clean work area.

White light and penetrating radiation must not be present in the darkroom because they can ruin any undeveloped film, including undeveloped radiographs that may be present. Darkrooms should be equipped with low-wattage lights with red filters (safelights). Safelights should be 0.9-1.2 m (3-4 ft) from any part of the darkroom where undeveloped film will be exposed. These lights will provide sufficient visibility for cutting film, loading and unloading film holders, and manual film processing.

A workbench for cutting film and loading cassettes and film holders should be located a considerable distance away from the processing machine or tanks. The bench must be kept clean and free from chemical spills and dirt that may scratch the film emulsion and it should be large enough to facilitate the workload. Sufficient storage areas for film, cassettes, screens, and film hangars must also be available and conducive to the workflow.

Manual Processing

For manual processing, the film is placed on a frame or film hanger where it is to remain until it has been dried after fixing. The film is then immersed in the developer for a time that depends on the temperature of the developer. Typically, development for 5 minutes at 20 °C

(68 °F) is used for manual processing, with shorter times at higher temperatures and longer times at lower temperatures. The film hanger should be tapped against the tank immediately after the film is completely submerged in developer to dislodge air bubbles adhering to the film. During development, the developer or the film must be agitated to allow fresh solution to contact the film emulsion frequently. If the film is developed without agitation, each area of the film will affect the development of the areas below it, causing uneven development and streaking.

At the end of the proper development time, the film is removed from the developer, allowed to drain for a few seconds, and then immersed and agitated for 30-60 s in a "stop bath" to halt the development process. Stop bath is an acid solution that neutralizes the residual developer in and on the film, and helps to prevent the film from streaking during fixing.

After the stop bath, film is placed in the fixer and agitated for 10-15 s. When the film is initially submerged in the fixer, it takes on a cloudy, milky-white appearance that should clear in about 1 minute if the fixer is at 20 °C (68 °F). After the film has cleared, it should remain in the fixer for an additional time equal to twice the time required for it to clear. If the film is not fixed thoroughly, it will discolor as it ages. Keep in mind that overfixing reduces the image contrast and density.

When fixing is complete, the film is washed in running water with a sufficient flow to rapidly carry away the fixer. The emulsion should remain in contact with constantly changing water that covers the top of the hanger. The washing time should be at least twice the fixing time to prevent later staining and fading of the image.

The film is then placed in a circulating warm-air drying cabinet, which should not exceed 49 °C (120 °F). Film should be removed from the dryer as soon as it is completely dry. A film is adequately dry when there is no moisture remaining underneath the hanger clips that could possibly drip down the film and cause streaking.

It is very important to control the temperature of the developer, stop-bath, fixer, and wash water during processing, and to time the exposure of the film to these liquids. The various processes occur at different speeds depending on the temperature. Temperatures of 18-24 °C (65-75 °F) are preferable. If higher temperatures are unavoidable, special precautions should be taken to avoid damaging the film. These may include use of special formulations of the chemicals or shorter processing times. Do NOT use ice in the chemical tanks because it will dilute the chemicals. It is also important to avoid having large temperature differences between the various liquids to avoid reticulation, frilling, or other damage to the film emulsion.

Automatic Processing

When a large number of films must be processed each day, a film processing machine or automatic film processor will provide economic advantages. Automated film processing reduces the manpower required in the darkroom, reduces the time required for processing, and aids in assuring consistent, high-quality processing.

Exposed film is placed directly into the processor without the need for film hangers. A series of rollers moves the film at a controlled speed through each step of the process. The processor maintains the chemicals at the proper temperatures, agitates and replenishes the solutions automatically, and dries the film. By using special chemicals and high temperatures, overall processing time can be greatly reduced. A total time as little as 5 minutes from dry-to-dry is possible with some loss of radiographic quality and a total time as little as 11 minutes can be attained with negligible loss of quality.

As with any machine, maintenance is important, but an hour of maintenance once a week by a trained technician is usually sufficient to avoid problems.

Exposure Techniques

Whenever possible, RT is performed with a technique in which the radiation passes through only one thickness or wall of an object. This single-wall RT requires that the source be located on one side of the object and the film on the other side, with no intervening material. Single-wall RT simplifies exposure calculations and interpretation of the resulting radiographic image. Nevertheless, complex shapes and variations in wall thickness may make it difficult to select the proper radiation energy, filters, screens, and film types, as well as film placement and other variables.

RT of hollow parts (i.e., pipes and pressure vessels) often cannot be performed with the single-wall technique because either the source or the film cannot be placed inside the object. When this occurs, it is necessary for the radiation to pass through two, and sometimes more, walls of the object. This is called double-wall RT. Most specifications detail special requirements for double-wall RT to assure that adequate sensitivity is obtained for both walls and to assist in the interpretation of the radiographs.

For large spherical or cylindrical objects where both the inside and outside surfaces are accessible, the panoramic technique is useful to reduce exposure time. Figures 1.11 and 1.12 show the general arrangement of the source and the film for a cylinder and a hemisphere. Although welds are discussed, the technique is equally useful for parts of similar shapes that are not welded. The major requirement is that the wall thickness be relatively constant for all films exposed at the same time and that the source-to-film distance be sufficient so that the geometric unsharpness will be satisfactory using the intended source location. The panoramic technique is also useful for the RT of many small similar parts at the same time, as shown in Figure 1.13.

Figure 1.12: Hemispherical orange-peel head exposure arrangement

Reprinted from ASNT's *NDT Handbook*, second edition: Volume 3, *Radiography and Radiation Testing*

Figure 1.11: Weld radiography of larger diameter pipes and pressure vessels

Reprinted from ASNT's *NDT Handbook*, second edition: Volume 3, *Radiography and Radiation Testing*

Figure 1.13: Panoramic exposure arrangement

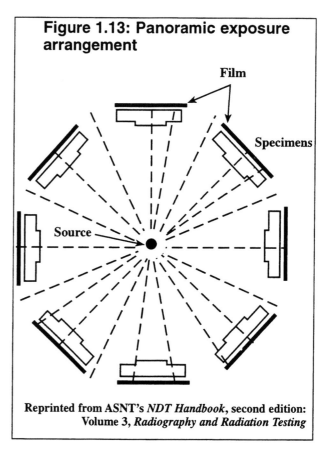

Reprinted from ASNT's *NDT Handbook*, second edition: Volume 3, *Radiography and Radiation Testing*

Recommended Reading

Subject	Reference*
image quality indicators	HB: 229-232; ASTM, ASME and MIL standards
identification markers	HB: 441; CT: 6-50
films and film handling	HB: 174-184
film graininess	HB: 182; CT: 4-8, 6-30
film selection	HB: 202; CT: 4-10
film handling and storage	HB: 237; CT: 4-16
film processing	HB: 300, 317, 327; CT: 4-10 to 4-11, 4-16
darkroom and equipment	HB: 313-314; CT: 4-16

*See *Introduction* for explanation of references.

Discontinuity Depth Determinations

Knowledge of the depth that a discontinuity lies below the surface of a part can be very useful to manufacturing personnel who must remove the discontinuity or to engineers who must determine its acceptability. There are several RT techniques that can be used for determining the depth of a discontinuity. These include stereoradiography and three parallax techniques, the rigid formula, the single-marker technique, and the double-marker technique.

The rigid formula and single marker techniques both rely on making two exposures on the same film. Therefore, they are useful only when the discontinuity is contrasty enough to be visible on a double-exposed radiograph (i.e., one film exposed twice, with each exposure being half of the normal exposure time). The double marker system does not have this limitation, which makes it the most generally applicable technique.

As shown in Figure 1.14, the double-marker technique uses lead markers placed on both the source and the film side of the object. Then, two normal exposures are made.

One film is exposed with the source in its normal position and the second is exposed with the source shifted an appreciable amount (20-30% of the source-to-object distance) parallel to the film plane. In the case of cracks, the source shift must also be parallel to the length of the crack. The images of the source side marker and the discontinuity will shift relative to the image of the film side marker. The shift of the source side marker will be proportional to the thickness of the object (the

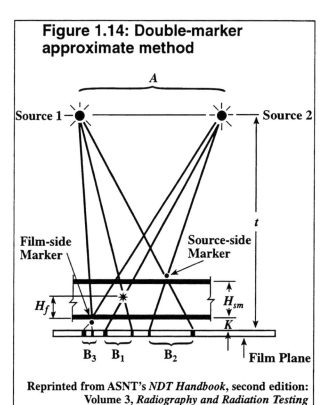

Figure 1.14: Double-marker approximate method

Reprinted from ASNT's *NDT Handbook*, second edition: Volume 3, *Radiography and Radiation Testing*

distance from the source side to the film side), while the shift of the discontinuity image is proportional to the distance from the discontinuity to the film side of the object. Knowing the thickness of the object, the distance of the discontinuity from the film side of the object is a simple algebraic calculation.

Interpretation and Evaluation of Radiographs

Interpretation of radiographs is the process of determining whether the radiograph is suitable for evaluating the condition of the object. The interpreter must be familiar with the requirements of the governing specification with regard to film identification and location markers, penetrameter size and placement, and film density.

Only personnel trained and experienced in RT should evaluate the condition of the object. The evaluator must be familiar with how the radiographic variables that were employed in making the radiograph may affect the image of the object and the image of the various kinds of discontinuities. The evaluator must be able to identify the images of various types of discontinuities and know which discontinuities are most likely to occur in a given portion of the object based on knowledge of how the object was made.

Visual Acuity and Dark Adaptation

The evaluator must have good visual acuity (the ability to see fine detail) and the ability to discern small changes in image density (i.e., low contrast images). For this reason, most specifications require radiographers, interpreters, and evaluators to have an annual examination for near-distance visual acuity, and some require an examination for brightness/contrast discrimination.

No matter how good an individual's visual acuity and brightness discrimination may be, the ability to see low contrast images is strongly affected by the light level to which the eyes have been exposed recently. For most interpretation and evaluation, it is sufficient to "dark adapt" one's eyes for at least 10 minutes by avoiding all white light. For the most critical work and for very low-contrast images, dark adaptation for at least 30 minutes is necessary.

Repeated adaptation can be avoided by wearing red goggles when exposed to white light.

Even when well dark-adapted, a useful procedure to improve the visibility of faint images is to move the radiograph back and forth during viewing, because indistinct objects are more easily seen when moving. For example, it easier to see an animal in the woods or fields when it is moving than when it is still. Viewing the film at an angle can also improve the visibility of faint images.

Viewing Conditions

In addition to dark adaptation, viewing conditions are very important to the interpretation of radiographs. The contrast sensitivity of the human eye is greatest when the surroundings are approximately the same brightness as the area of interest. Therefore, to avoid loss of dark adaptation and to provide the best conditions for seeing, the film illuminator must be masked to prevent bright light from escaping around the edges of the radiograph. For the same reason, when there are light areas in a dark radiograph, a spot illuminator should be used to prevent glare from the light areas. Lighting in the viewing area should be subdued and the viewer placed to minimize unwanted to reflections from other light sources.

Film Density Measurement

The area of the radiograph under evaluation must be within a specific density range. Most specifications require densities between 1.5 and 4.0. However, some specifications allow densities above 4.0 provided that the that the illuminator is capable of supplying sufficient light to penetrate these densities.

Film density should be measured with a densitometer, an instrument that compares the intensity of the light transmitted through the film to the light intensity incident on the other side of the film. The densitometer should be allowed to warm up for several minutes and then be calibrated using a calibrated density strip. A series of density readings should be taken in the density range that is required for the radiographs that are to be to be evaluated.

Identifying Indications

Indications on a radiograph may be relevant or nonrelevant. Nonrelevant indications are those that were produced by:
1. features of the object that are intended to be present, such as its shape, including holes, ridges, or steps in the object, or
2. errors in RT techniques, such as water spots, scratches, or pressure marks on the film.

Nonrelevant indications caused by film handling or processing errors are often called film artifacts. A common method of aiding in identification of artifacts is to simultaneously expose two films of the same type in the film holder to produce two radiographs that are nominally identical. If the same indication is present on both radiographs, it is either an intended feature of the object or a discontinuity. Despite such aids, there is no substitute for experience in determining whether indications are relevant or not. Table 1.5 lists the common types of artifacts and their causes.

Relevant indications are those produced by unintentional conditions in the object, usually discontinuities such as voids, inclusions of foreign material, or cracks.

Sources of Discontinuities

Discontinuities may be created at any stage during the life of a metal or other material – from its initial formation through the end of its service life. They are often roughly classified as inherent discontinuities, processing discontinuities, or service discontinuities, depending on the stage during which they were created.

Inherent Discontinuities

Inherent discontinuities are those that were created during the smelting, refining, pouring, and solidification of the metal into ingots, billets, or slabs. They usually result from entrapment of foreign material such as absorbed gases, oxides, or sulfides, although some discontinuities such as pipe are a result of mold design or pouring practices.

Processing Discontinuities

Processing discontinuities include those produced during processes that modify the shape and/or properties of the raw material, such as forging, rolling, heat treating,

Table 1.5: Film artifacts

Type of Artifact	Cause
Crimps (fold/bend)	Careless film handling technique
Static (tree branches)	Rough/rapid handling of film loading or unloading holder
Scratches	Handling, processor rollers (visible on film surface viewed with light at an angle)
Pressure mark	Heavy object placed on loaded film holder or bending cassette
Screen marks	Blemishes/contamination on screens (produce light image on film)
Light leak (light streaking)	Tears in cassette prior to developing
Fog (mottled film)	Exposure of unprocessed films to temperature, chemicals, atmosphere, radiation, or light
Light leak (dark streaking)	Tears in cassette prior to radiographic exposure
Chemical spots/streaks	Manual process chemical splashing, no agitation in developer, no stop bath used
Processor pressure marks	Dark areas caused by roller contamination or improper clearance
Miscellaneous processing	Cross contamination of chemicals, air bells (cause false indications)

machining, or grinding. Stresses and deformation are major causes of processing discontinuities such as cracks, bursts, laps, and seams, though environmental factors such as temperature may also be a cause. Discontinuities in castings and welds also are usually considered to be processing discontinuities, although many of them are similar to inherent discontinuities because the processes involve the melting and solidification of metal.

Service Discontinuities

Service discontinuities are the result of stresses and environments imposed on the part during service. Cracking may result from processes such as fatigue, creep, thermal shock, or stress-corrosion. Pitting or general metal loss may be caused by corrosion, erosion, wear, or fretting. Service discontinuities are usually open to the surface and, therefore, more economically detected by surface methods of NDT such as liquid penetrant testing, eddy current testing, or magnetic particle testing. However, the affected surface may be internal to the part (i.e., the ID of a pipe or a rivet hole) or in the inner surfaces of a sandwich construction, and thus not accessible for surface tests. RT may then be the best method of detection.

Identifying Discontinuities

It is very important to manufacturing and engineering personnel to know what kind of discontinuity caused a given NDT indication. Manufacturing personnel can use the information to try to prevent the recurrence of such discontinuities, while engineers are concerned because some discontinuities have more effect than others do on the usefulness and service life of an engineered structure.

Because the radiographic appearance of a discontinuity depends on shadow formation, the technique variables used in producing the radiograph have a significant effect on the appearance of the discontinuities. Therefore, accurate evaluation of radiographs depends on the training, experience, and skill of the evaluator. Accurate evaluation skills can only be learned by extensive practice combined with a thorough knowledge of radiographic principles.

In addition to the influences of radiographic variables, there may be wide variations in the size and shape of any given type of discontinuity, and discontinuities may be oriented in various ways relative to the radiation beam. As a result of these complexities, verbal descriptions of the radiographic appearance of various types of discontinuities are at best ambiguous and may be confusing. The proper way to become familiar with the radiographic appearance of discontinuities is to study the reference radiographs published by ASTM and other reputable technical organizations and gain extensive experience.

Recommended Reading

Subject	Reference*
discontinuity depth determinations	HB: 808-816
radiographic evaluation and interpretation	HB: 337; CT: 1-6
visual acuity and dark adaptation	HB: 382, 384
viewing conditions	HB: 384
radiographic film density	HB: 179, 384; CT: 4-6
identifying indications	HB: 378; CT: 7-4

*See *Introduction* for explanation of references.

Radiographic Inspection Documents

In addition to the radiographs themselves, records should be made of the RT technique used, the results of the interpretation, and the identity of the individuals involved. Specifications and contracts usually impose detailed lists regarding such records, but as a minimum they should include the following:

1. identification of the part or parts radiographed, including the drawing and serial numbers,
2. descriptive name(s) for the part(s),
3. governing contract or specification, if any,
4. type of material,
5. surface condition of part when radiographed,
6. RT technique used for each exposure, including:
    a. identification markers,
    b. type and identification of radiation source,
    c. thickness of material,
    d. source-to-film distance,
    e. penetrameters used,
    f. film type, size, and quantity,
    g. size and type of screens, filters, masking, etc., if any
    h. sketch or reference to:
       (1) location marker placement,
       (2) For multi-exposure parts, area(s) radiographed, and
       (3) arrangement of source, part and film, including radiation beam direction.
7. evaluation results for each radiograph, and
8. signature(s) of the radiographer and the evaluator.

For repetitive work on identical or similar parts, it is useful to keep a permanent record of the technique used. This will not only reduce the time needed subsequently to determine correct layout and technique, but it will allow these details to be documented for each part by reference to the record, rather than repeatedly recording it for each part. This is often done by means of serially numbered shooting sketches or technique sheets kept in a permanent file.

Radiation Safety

When the human body absorbs radiation, the ionization that is produced damages the body. If the amount of radiation absorbed is small and it is spread over considerable time, the damage may be temporary because the body is able to repair it, the same way it repairs bruises, scrapes, or small cuts. However, if a large amount of radiation is absorbed in a short time, the damage may be permanent because it is too great to be repaired by the body. In extreme cases, death may result. Consequently, persons working with radiation must thoroughly understand the safety issues and what safety precautions must be taken.

In regard to safety precautions, two particular problems with radiation are that:

1. X-rays, gamma rays, and other ionizing radiation cannot be detected by any of the human senses, and
2. the damage may not be apparent immediately; therefore, it is extremely important to follow all radiation safety rules.

Safety practices are based on known medical facts about how the body is affected by radiation. The legal limit of radiation that an individual may be exposed to, and other requirements, are set by governmental bodies such as the United States Nuclear Regulatory Commission (USNRC) and various state bodies. Radiation detection instruments must be used whenever a radiographic exposure is made regardless of the radiation source. Dosimeters and film badges must be worn by anyone working within a radiation area.

Fixed radiation facilities must be surveyed before the first use, provided with permanent monitoring instruments and alarms, and periodically surveyed for radiation safety. Field radiation work areas must have clearly marked boundaries, and be thoroughly monitored during operation. The size of fieldwork areas should be minimized as much as possible by use of devices such as portable shields and collimators to reduce the potential for accidental exposure of radiographers and other personnel.

Because some radiation regulations vary from state to state, radiographers must be familiar with the laws governing RT in the state where they are working.

Recommended Reading

Subject	Reference*
radiographic inspection documents	HB: 381, 393
radiation safety	HB: 137; CT: 5-3

*See *Introduction* for explanation of references.

Chapter 1
Questions

1. Charged particles passing through matter lose energy primarily because of:

   a. scatter radiation.
   b. ionization.
   c. secondary scatter.
   d. charged electrons.

   HB:81; CT:2-15

2. When electrons of many different energies strike a target, a continuous spectrum of X-rays is generated. These X-rays are known as:

   a. bremsstrahlung.
   b. Compton X-rays.
   c. scattering.
   d. slow electron emission.

   HB:82; CT:2-9

3. To produce an exposure equivalent to 5 mA at 305 mm (12 in.), what current is required if the source-to-film distance is changed to 609 mm (24 in.) and the exposure time is kept the same?

   a. 4 mA.
   b. 10 mA.
   c. 20 mA.
   d. 40 mA.

   HB:198; CT:2-11

4. A source-to-film distance of 762 mm (30 in.) is changed to 609 mm (24 in.). What exposure time would be required if the original exposure time was 10 minutes?

   a. 2.5 minutes.
   b. 5.0 minutes.
   c. 6.4 minutes.
   d. 15.2 minutes.

   HB:199

5. What is the intensity at 9 m (30 ft) for an unshielded X-ray source that is 232 mC/kg/hr (900 mR/hr) at 3 m (10 ft)?

   a. 26 mC/kg/hr (100 mR/hr).
   b. 64 mC/kg/hr (250 mR/hr).
   c. 116 mC/kg/hr (450 mR/hr).
   d. 258 mC/kg/hr (1000 mR/hr).

   HB:198

6. Geometric unsharpness can be reduced by using a:

   a. larger focal spot size.
   b. smaller focal spot size.
   c. shorter source-to-film distance.
   d. longer object-to-film distance.

   CT:6-25

7. It is required that the radiographs of a part 102 mm (4 in.) thick must have a geometric unsharpness no larger than 0.4 mm (0.015 in.). If the maximum projected dimension of your radiation source is 2.5 mm (0.1 in.), what minimum source-to-film distance must you use to satisfy the unsharpness requirement?

   a. 66 cm (26 in.).
   b. 76 cm (30 in.).
   c. 91 cm (36 in.).
   d. 102 cm (40 in.).

   CT:6-29

8. All other factors being the same, radiographic sharpness or definition is improved by using:

   a. slower film.
   b. faster film.
   c. film with a larger grain size.
   d. slower film with fluorescent screens.

   CT:6-26

9. Other factors being the same, radiographic contrast is improved by:

   a. raising the kilovoltage and lowering the current.
   b. lowering the kilovoltage.
   c. using faster film.
   d. adjusting the exposure to produce the minimum film density allowed.

   HB:180; CT:6-27

10. The minimum source-to-film distance needed to produce acceptable radiographs depends on the focal spot (source) size, the maximum allowable unsharpness, and the:

    a. type of film.
    b. density of the object.
    c. object-to-detector distance.
    d. atomic number of the object material.

    CT:6-28

11. A significant difference between automatic and manual processing is that:

    a. the chemistry in automatic processors is more uniform.
    b. manual processing is faster.
    c. automatic processing is more reliable and cost effective.
    d. developer time is not critical with automatic processing.

    CT:4-14

12. The darkroom safelights for RT should have colored filters and frosted white bulbs, and be placed 1-1.2 m (3-4 ft) from the darkroom work surfaces. What should be the wattage of the bulbs?

    a. 7.5-15 W.
    b. 15-20 W.
    c. 20-30 W.
    d. 30-40 W.

    HB:314

13. Because white light bulbs are used in darkroom safelights, the light must be filtered with a:

    a. blue/green filter.
    b. red/amber filter.
    c. dark filter of any color.
    d. yellow/green filter.

    HB:313

14. The loading bench in a darkroom should be:

    a. next to the film processing tanks or machine.
    b. away from the film processing tanks or machine.
    c. near the entrance for convenience in passing films in and out of the darkroom.
    d. near the vent fan to assure a good rate of air flow over the film as it is loaded in the cassettes or film holders.

    HB:312

15. Static marks in radiographs may result from:

    a. using contaminated cotton gloves.
    b. using noninterleaved film.
    c. sliding film rapidly out of the film holder.
    d. loading several films into one film holder.

    HB:315

16. When using strips of roll film, they must be sealed with dark tape to prevent:

    a. light spots on the film.
    b. exposure to light.
    c. lighter densities on developed films.
    d. burn out of lead identification numbers.

    HB:234

17. Unexposed film should be stored in an area protected from heat, humidity, light, and:

    a. penetrating radiation.
    b. electrical fields.
    c. magnetic fields.
    d. red light.

    HB:235

18. The liquids used for manual processing should include:

    a. developer, stop bath, and water.
    b. developer, fixer, stop bath, and wetting agent.
    c. fixer, stop bath, and water.
    d. developer, stop bath, fixer, and water.

    CT:4-11

19. Adding chemicals to restore the activity of a developer solution during normal use is known as:

    a. reactivation.
    b. replenishment.
    c. restoration.
    d. renovation.

    CT:4-11

20. The basic purpose of a stop bath is to:

    a. cause development to cease.
    b. speed up the fixing process.
    c. enhance the alkalinity of the developer that on the film.
    d. prevent excessive fixation.

    CT:4-12

21. The basic purpose of the fixer is to:

    a. soften the film emulsion.
    b. remove the unexposed silver halides.
    c. reduce the alkalinity of the developer.
    d. neutralize the developer acids.

    CT:4-12

22. Other factors being equal, processing film in solutions that are too warm may result in:

    a. lower densities.
    b. mottling.
    c. frilling or loosening of the emulsion.
    d. uneven densities.

    CT:4-12-15

23. Washing film in cold water (less than 10 °C, (50 °F) during manual film processing results in:

    a. greater washing action.
    b. little washing action taking place.
    c. uniform drawing of water from the film.
    d. improved results from the film drying process.

    CT:4-13

24. Using a wetting agent in manual film processing will assist in:

    a. the developing stage of the process.
    b. hardening the film emulsion.
    c. reducing water marks and streaks.
    d. controlling film density.

    CT:4-14

25. For best results, film should be dried:

    a. in ambient air.
    b. by an oscillating fan.
    c. in a dryer for a minimum of 2 hours.
    d. in a warm, filtered air dryer.

    CT:4-15

26. The major advantages of automatic film processing are:

    a. fewer spills and splashes in the darkroom.
    b. improved sensitivity of the radiographs.
    c. reduced cost and time for processing.
    d. speed, consistency and efficiency.

    HB:348

27. Which of the following should be done to preserve radiographs during long-term storage?

    a. Seal the radiographs in plastic envelopes.
    b. Keep the radiographs away from bright light or sunlight.
    c. Do not use interleaving paper between the radiographs.
    d. Store radiographs horizontally, on top of each other.

    HB:238

28. Excessive density in a radiographic image may be a result of:

    a. insufficient fixing.
    b. excessive exposure time.
    c. the use of fine grain film.
    d. the use of an X-ray energy greater than 200 kV.

    HB:182

29. The difference in density between different parts of a radiograph is called:

    a. radiographic contrast.
    b. radiographic latitude.
    c. gamma of the film.
    d. subject contrast.

    HB:182

30. Overall image quality of a radiograph is determined by its radiographic contrast and:

    a. definition.
    b. density.
    c. sensitivity.
    d. latitude.

    HB:273

31. In manual processing, low radiographic density may be due to:

    a. high developer temperature.
    b. high developer concentration.
    c. weak developer solution.
    d. over-replenishing.

    CT:4-10

32. Over-developing may result in:

    a. streaking.
    b. fogging.
    c. spotting.
    d. lower density.

    CT:4-11

33. Unwanted marks and images that are produced during the processing of a radiograph are known as:

    a. anomalies.
    b. irregularities.
    c. artifacts.
    d. relevant indications.

    HB:398

34. A densitometer is an instrument that is used to measure:

    a. the X-ray density of an object.
    b. the depth of color of an object.
    c. the physical density of an object.
    d. the transmittance or reflectance of an object.

    HB:388

35. Which of the following factors is most important in assuring that satisfactory radiographs can be stored for years without becoming useless?

    a. Development beyond 5 minutes at 20 °C (68 °F).
    b. Using low pH stop bath.
    c. Thorough washing to remove all the thiosulfate.
    d. Fixing for at least three times the normal clearing time.

    HB:310

36. Densitometers should be calibrated before use by:

    a. taking a series of readings from a calibrated density strip.
    b. using production radiographs with known densities.
    c. measuring the reflected light from a radiograph.
    d. using a photometer.

    HB:389

37. Grinding cracks are not readily discernible by RT because:

    a. they are in the wrong direction.
    b. they are too tight and shallow.
    c. products are usually inspected prior to grinding operations.
    d. there produce no difference in the material density.

    CT:7-15

38. Stress corrosion cracking may sometimes be detected with RT when using high definition and high contrast techniques and when:

    a. a major dimension of the corrosion is perpendicular to the X-ray beam.
    b. a major dimension of the corrosion is parallel to the X-ray beam.
    c. for thin parts, the radiation source is a betatron.
    d. the corrosion was not caused by nuclear reactor water.
    CT:7-60

39. Discontinuities caused by fatigue and/or corrosion are categorized as:

    a. processing discontinuities.
    b. inherent discontinuities.
    c. service discontinuities.
    d. metallurgical discontinuities.
    CT:7-6

40. Heat-affected zone cracking is not easily detected with RT because:

    a. of the orientation of the cracks.
    b. the cracks are too tight to be detected.
    c. the cracks are too shallow to be detected.
    d. the cracks are very irregular in shape.
    CT:7-18

41. Inherent discontinuities are those found in:

    a. forgings.
    b. plate.
    c. ferrous metals.
    d. ingots.
    CT:7-3

42. Nonmetallic inclusions in a casting appear on a radiograph as:

    a. nearly round dark spots
    b. clustered, elliptical dark spots.
    c. irregularly-shaped dark spots.
    d. smooth, long, narrow dark lines.
    HB:473

43. Which of the following discontinuities is not related to the casting process?

    a. Cold shuts.
    b. Hot tears.
    c. Laminations.
    d. Sponge shrinkage.
    HB:472

44. Radiographs of metals such as brass, austenitic stainless steel, and some nickel and cobalt-base alloys sometimes show "diffraction mottling." Diffraction mottling can be distinguished from discontinuities by re-radiographing using:

    a. slightly lower radiation energy.
    b. slower speed film.
    c. a slightly different radiation angle relative to the object.
    d. an X-ray diffraction machine.
    HB:211

45. A straight, smooth, linear indication in the center of a weld joint is most likely caused by:

    a. lack of fusion.
    b. lack of penetration.
    c. undercut.
    d. centerline segregation.
    HB:455

46. What is the HVL thickness of aluminum for a radiation beam that requires 10 minutes exposure for 25 mm (1 in.) of aluminum and 20 minutes exposure for 38 mm (1.5 in.) of aluminum, all other factors being the same?

    a. 13 mm (0.5 in.).
    b. 25 mm (1 in.).
    c. 38 mm (1.5 in.).
    d. Not enough data are given.

47. A groove melted in the base metal at the edge of weld is called:

    a. lack of penetration.
    b. lack of fusion.
    c. undercut.
    d. melt-back.
    HB:475

48. The density differential between two areas of a radiograph is radiographic:

    a. sensitivity.
    b. definition.
    c. contrast.
    d. filtration.

    HB:228

49. The ratio of radiation intensities transmitted by various sections of a part as a result of thickness changes, is:

    a. subject contrast.
    b. radiographic contrast.
    c. sensitivity.
    d. film gradient.

    HB:227

50. What source-to-film distance is required to produce a maximum geometric unsharpness of 0.5 mm (0.02 in.) for an object 142 mm (5.6 in.) thick using a 5 mm (0.2 in.) focal spot (source) size?

    a. 142 cm (56 in.).
    b. 156 cm (62 in.).
    c. 226 cm (89 in.).
    d. 274 cm (108 in.).

    HB:196

51. The penetrating power of an X-ray beam is a function of its:

    a. wavelength.
    b. source-to-object distance.
    c. angle of incidence.
    d. tube current.

    HB:71

52. As the effective energy of the radiation increases:

    a. film graininess decreases.
    b. film graininess increases.
    c. radiographic definition increases.
    d. film speed decreases.

    HB:182

53. Devices used to assure that radiographs meet the desired image quality level are called:

    a. penetrameters.
    b. location markers.
    c. diaphragms.
    d. shims.

    HB:439

54. A front filter at the film or at the source, or multiple-films in the same holder, are techniques that are used to compensate for:

    a. excessive subject contrast.
    b. low density in thin areas of the object.
    c. poor definition.
    d. low subject contrast.

    HB:209, 512; CT:6-12

55. The energy level of X-rays is controlled by the:

    a. size of the target.
    b. filament voltage.
    c. filament current.
    d. voltage applied between the cathode and anode.

    CT:2-10

56. The maximum tube current of an X-ray tube is determined by the:

    a. composition of the cathode.
    b. distance from the anode to the cathode.
    c. target material, size of the focal spot, and efficiency of the cooling system.
    d. material used to construct the vacuum envelope.

    HB:192; CT:3-12

57. At 100 kV, the radiographic equivalence factor for aluminum is 1.0 and for magnesium is 0.6. Approximately what thickness of aluminum is 51 mm (2 in.) of magnesium equivalent to:

    a. 17 mm (0.7 in.).
    b. 23 mm (0.9 in.).
    c. 31 mm (1.2 in.).
    d. 85 mm (3.3 in.).

    HB:203

58. If the source-to-film distance for a technique is modified from 76 cm (30 in.) to 114 cm (45 in.), the exposure increases by a factor of:

 a. 0.44.
 b. 0.67.
 c. 1.50.
 d. 2.25.

HB:197

59. A 3.7 TBq (100 Ci) source of Ir-192 is to be used to radiograph 51 mm (2 in.) of steel. If the exposure factor for 51 mm (2 in.) of steel to produce a film density of 1.5 is 0.03 TBq-min/mm² (0.7 Ci-min/in.²), how many minutes exposure will required with this source using a (30 in.) source-to-film distance?

 a. 2.3 minutes.
 b. 3.2 minutes.
 c. 6.3 minutes.
 d. 7.2 minutes.

HB:228

60. Crimp marks (small, arc-shaped marks) that are lighter than the surrounding image on a radiograph are the result of:

 a. poor handling before exposure.
 b. poor handling after exposure.
 c. arc-shaped scratches on the intensifying screen.
 d. splashing fixer on the film before developing.

HB:397

61. You have an Ir-192 source whose strength was 3.7 TBq (100 Ci) on July 1 this year. Approximately what will be its strength on September 13 this year if the half life of Ir-192 is 75 days?

 a. 1 TBq (30 Ci).
 b. 1.9 TBq (50 Ci).
 c. 2.6 TBq (70 Ci).
 d. 3.3 TBq (90 Ci).

62. After three half-lives have elapsed, what will be the strength of an isotope source relative to its original strength?

 a. 25% of the original.
 b. 33% of the original.
 c. 12.5% of the original.
 d. 11% of the original.

CT:2-21

63. For a given film, 1.0 is the log relative exposure for a density of 0.4 and 2.0 is the log relative exposure for a density of 2.5. Using this film and an exposure of 1 mAm, you have made a radiograph with a density of 0.4. What exposure must you give to obtain a radiograph with a density of 2.5 on this film, all other factors being the same?

 a. 2 mAm.
 b. 5 mAm.
 c. 10 mAm.
 d. 20 mAm.

64. What radioisotope has a gamma ray energy emission that is approximately equal to 320 kV radiation from an X-ray tube?

 a. Co-60.
 b. Ir-192.
 c. Cs-137.
 d. Ur-238.

CT:2-22

65. The specific activity of a gamma ray source is important because it determines:

 a. the size of source required for a given activity.
 b. the energy of the radiation that will be emitted.
 c. the rate at which the specific isotope decays.
 d. which type of radiation (alpha, beta or gamma) that it will emit.

CT:2-21

Chapter 1
Answers

1.	b	18.	d	35.	c	52.	b
2.	a	19.	b	36.	a	53.	a
3.	c	20.	a	37.	b	54.	a
4.	c	21.	b	38.	b	55.	d
5.	a	22.	c	39.	c	56.	b
6.	b	23.	b	40.	a	57.	c
7.	b	24.	c	41.	d	58.	d
8.	a	25.	d	42.	c	59.	c
9.	b	26.	d	43.	c	60.	a
10.	c	27.	b	44.	c	61.	b
11.	c	28.	b	45.	b	62.	c
12.	a	29.	a	46.	a	63.	c
13.	b	30.	a	47.	c	64.	b
14.	b	31.	c	48.	c	65.	a
15.	c	32.	b	49.	a		
16.	b	33.	c	50.	b		
17.	a	34.	d	51.	a		

Chapter 2
Procedure Comprehension and Instruction Preparation (PCIP) Examination

Overview of the PCIP Examination Process

Candidates for ASNT Central Certification Program (ACCP) Level II certification come from different industries with differing experiences and unique terminology regarding what constitutes written directions that guide an individual in performing a nondestructive test. Some of the terms that may be familiar are procedure, instruction, job card, technique, and practice; often words like "specific" and "general" appear with these terms to further define the level of detail these written directions convey. To maintain consistency within the ACCP examination process, ASNT adopted a particular set of terms. The following description of these terms is presented to familiarize the candidate with the ACCP examination terminology.

Definitions of PCIP Examination Terminology

Specification

Figure 2.1 shows the hierarchy of ACCP documents. It begins with a specification that is usually received from a client and it often references a national standard or code. A specification is a list of requirements and qualifying conditions for the test object and the NDT that is to be performed on the test object. For the purposes of the ACCP, the specification is not significant; however, the Level II candidate should understand what a specification is in relationship to the other ACCP documents.

Procedure

A procedure is a written description that establishes minimum requirements for performing an NDT method on any object and it is written in accordance with established standards, codes, or specifications. A procedure is written by certified Level III personnel in that NDT method for which the procedure applies. Its primary purpose is for use in developing NDT instructions. A procedure may be used to perform an NDT technique; however, due to its rather general nature concerning requirements, conditions, and limitations for an NDT method, it demands greater knowledge, some interpretation, and judgment on the part of the inspector in order to develop a specific NDT technique and apply it to a specific part.

The PCIP portion of the ACCP Level II General Written Examination requires a candidate to read and comprehend a procedure. Because Level III personnel prepare procedures

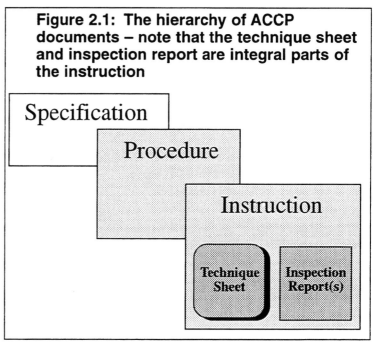

Figure 2.1: The hierarchy of ACCP documents – note that the technique sheet and inspection report are integral parts of the instruction

within the ACCP, each procedure will be different. However, a procedure will generally have a structure similar to the following:
1. scope,
2. references,
3. personnel certification requirements,
4. equipment and materials requirements,
5. calibration and verification requirements,
6. part preparation requirements,
7. test sequence requirements,
8. interpretation and evaluation requirements,
9. documentation requirements,
10. instruction and technique sheet requirements, and
11. post-inspection requirements.

Instruction

An instruction is a description of the steps to be followed when performing an NDT technique on a specific part or a set of similar parts, developed in conformance with a procedure. An instruction is written by personnel certified to at least Level II in that NDT method for which the instruction applies. Its primary purpose is to provide sufficient direction to enable Level I and Level II personnel, as applicable, to perform an NDT technique that yields consistent and repeatable testing results. While a procedure is generally written around an NDT method, an instruction provides specific details concerning a particular technique for a given NDT method. An instruction is also valuable because it reduces or eliminates different inspector interpretations of a procedure or divergent NDT techniques performed on the same test object. A technique sheet and an inspection report are integral parts of an instruction.

An instruction usually contains a scope, information about personnel/equipment and material/calibration, and the test sequence.

The scope identifies the procedure (the minimum requirements) to which the instruction applies as well as any conditions that apply before the inspection is to be conducted. For example, a scope might indicate that heat treating, forming, or hydro-testing processes be performed before the inspection and that preapproval of approaches and agreements to exclusions should be negotiated.

The title or the scope should indicate the NDT method and technique of interest.

The personnel/equipment and materials/calibration categories are addressed under separate headings or within other areas of the instruction such as the test sequence or the technique sheet as appropriate. While some techniques are relatively simple and require less headings, others are complex and require a greater number of headings to efficiently inform the inspector of the necessary details.

The test sequence is the step-by-step listing of actions that are to be taken in the order in which they are to be carried out. Generally, each step is an action and it should begin with a verb such as "prepare," "conduct," "verify," "interpret," or "report." The candidate should be referred to the applicable technique sheet that contains quantitative or graphical information that will aid him/her in performing the inspection.

Technique Sheet

A technique sheet identifies all of the parameters required to set up the technique for which the instruction applies. A technique sheet should have a diagram that shows the test object and pertinent areas of inspection. It may be necessary to present more than one view to adequately inform other inspectors of the set-up. The part's identification and acceptance criteria are also present on the technique sheet.

Included in a technique sheet are:
1. equipment settings and materials [e.g., milliampere-minutes, becquerel (curie) minutes, film size and type, source-to-film distance];
2. part identification, material, and other unique information;
3. areas of inspection interest;
4. acceptance criteria; and
5. identification and certification of the individual who developed and wrote the instruction, the date it was written and revised, and it should be cross referenced to the applicable instruction.

Further, a technique sheet accompanies any instruction and identifies all parameters required to set up the technique for which the instruction applies. Technique parameter requirements are to be incorporated into a technique sheet that, when used by a qualified Level II or Level I, will address the operational parameters necessary to conduct the inspection. The text of the instruction, being broader-based, identifies all of the steps to be taken including nonquantitative matters such as the general administration of the test and the recording of test results. The generally qualitative text of the instruction follows the generally quantitative, technical requirements listed in the technique sheet. Specific quantitative values, or values of limited extent, are to be used in the technique sheet (as opposed to the broad allowable ranges commonly found in standards and procedures).

The contents of a typical radiographic technique sheet are shown in Table 2.1. Details within each of the categories would vary considerably based on the product form (weld, casting, electronic assembly, composite), material (steel, aluminum, nonmetallic), the size and shape of the part ('T' section, flat plate, cylinder), the type of source (X-ray, Ir-192, Co-60), and acceptance criteria (applicable code, standard, or intended use). See Figure 2.2 (page 47) for a sample technique sheet.

Inspection Report

An inspection report usually identifies all detected discontinuity indications and whether they have been interpreted as relevant or nonrelevant. Identified relevant indications are evaluated as acceptable or rejectable in accordance with the part's acceptance criteria. The inspection report should include the location, direction, and dimensions of all rejectable indications detected on the part(s), and the identification of the rejectable part(s). The inspection report may include the location, direction, and dimensions of all acceptable indications detected on the part(s).

Forms used for reporting inspection results, logging actions, and other record keeping and administrative actions are to be addressed in the instruction. The forms to be used in recording the results of the inspection should be identified and included as attachments to the instruction, because they are required actions while performing NDT to an instruction. The contents of a typical radiographic inspection report are shown in Table 2.2. Please remember that the data on an inspection report is specific to a single inspection.

While a procedure is generally written around an NDT method, an instruction

Table 2.1: Elements of a typical radiographic technique sheet

Categories	Relevant Information (as applicable)
Test part	identification, model, material, product form
Authorization	contract number, drawing number, instruction number
Equipment and materials	radiation source type, size, and activity or settings, film type(s) and size(s), lead screens and sizes, backing lead, source to film distance, IQI(s) and shim type and size
Sketch of exposure set-up	location of location markers, shims, IQI(s), film identification, orientation of the source with respect to the part, area of interest, acceptance criteria zones
Acceptance criteria	relevant and nonrelevant indications, acceptable/rejectable: length, diameter, grouping of indications
Film development	times and temperatures
Author's identification	name, level, date, revision

Table 2.2: Elements of a typical radiographic inspection report

Categories	Relevant Information (as applicable)
Test part	identification, model, material, product form
Authorization	purchase order number, job number, instruction number, technique number
Equipment and materials used (when not accompanied by technique sheet or when different than technique sheet)	radiation source type, size, and activity or settings, film type(s) and size(s), lead screens and sizes, backing lead, source to film distance, IQI(s) and shim type and size
Coverage	areas inspected, areas that could not be inspected, partial or 100% inspection
Test results	rejectable and reportable indications, sizes, locations
Sketch of results (when applicable)	location (including depth as applicable) where repair or sectioning for investigation may be necessary
Inspector	name, level, date

provides specific details concerning a particular technique for a given NDT method.

For each instruction there is an accompanying technique sheet and inspection report. The instruction identifies and details all information that is not on the technique sheet. The instruction's quantitative details are contained in the technique sheet and not in the instruction's general text. See Figure 2.3 (page 48) for a sample inspection report.

Figure 2.4 (page 49) summarizes the contents of a specification, a procedure, an instruction sheet, a technique sheet, and an inspection report.

Format of the PCIP Examination

The PCIP portion of the ACCP Level II General Examination examines the candidate's ability to comprehend a procedure and requires the candidate to demonstrate the ability to prepare an instruction for NDT personnel with respect to a given procedure and technique parameters. However, during the examination the candidate is not required to write an instruction from beginning to end. Instead, a technique sheet and an inspection report form are provided so the candidate can extract information from the given procedure, make technique judgments, and then complete the appropriate areas on the technique sheet and inspection report. The candidate's comprehension of the given procedure and his/her ability to prepare instructions is further assessed by approximately 30 multiple-choice questions (these multiple-choice questions are in addition to the minimum of 60 multiple-choice General Examination questions that assess the candidate's knowledge of fundamentals and principles in the method). Figure 2.5 (page 50) graphically presents the PCIP portion of the Level II General

Figure 2.2: Sample technique sheet

Technique Sheet

Part: pipe weld	**Material:** carbon steel: 152 mm (6 in.) diameter - sch. 40	**Technique:** X542984	
Exposure: single wall ☐ double wall ☒		**Pb Screens:** front/back: 0.3/0.3 mm (0.01/0.01 in.)	
Shim(s): 2.4 and 1.6 mm (0.09 and 0.06 in.)	**Film Type:** double loaded x and y	**Film Size:** 114 × 254 mm (4.5 × 10 in.)	
IQI(s): ASTM #10 (S/S)	**Quality Level:** 2-2T	**Source:** Ir-192	**SFD:** 165 mm (6.5 in.)
Acceptance Criteria: AX541984	$U_{gm\,ax}$: 0.5 mm (0.02 in.)	**Exposure Time:** 0.4 TBq*min (12 Ci*min)	
Manual Development:	**Time:** 5 min.	**Temperature:** 20 °C (68 °F)	

Exposure Sketch

Pb B

ABC company, date, weld number, project number

Symbols

- ▭• = Penetrameter
- Location Markers = 0, 1, 2 ...
- ✶ = Source
- B = Lead Letter B
- ▓ = Backing Lead
- ── = Film Holder w/Screens
- ▨ = Identification Pad Location

Prepared By: *Jane Doe* **Level:** II **Date:** *01-01-99*

Figure 2.3: Sample inspection report

Inspection Report

Part: pipe weld	Surface Condition: as welded	Technique Number: X541984
ID Number: 22	Project: XYZ	Other: N/A
Quality Level: 2-2T	Acceptance Criteria: AX541984	

Film Interval No.	Accept (✓)	Reject (X)	Burn-through	Crack	Excess Penetration	Excess Reinforcement	High/low	Incomplete Penetration	Internal Concavity	Lack of Fusion	Porosity	Slag	Tungsten	Undercut	Remarks/Comments 1. If rejectable, state the particular discontinuity and reason for rejecting it. 2. Record film artifacts and radiographic quality issues here.
#1 0-1		X		X											
#2 0-1		X						X							
#3 2-3	✓												X		
#4 2-3		X									X				
#5 3-0		X													*film scratch*
#6															
#7															
#8															
#9															
#10															

Sketch

Not Required

Inspector: *John Doe*   Level: II   Date: *12-12-99*

Figure 2.4: Contents of a *Specification*, a *Procedure*, an *Instruction Sheet*, a *Technique Sheet*, and an *Inspection Report*

Specification
- What and where to inspect
- How to inspect (usually to a referenced code or standard)
- Acceptance criteria (if not in the reference materials)
- What and how to report results
- Who can inspect (certification)

Procedure
- Minimum requirements for the NDT method in general, including technique limitations:
  - Scope
  - References
  - Personnel certification requirements
  - Equipment and materials requirements
  - Calibration and verification requirements
  - Part preparation requirements
  - Test sequence requirements
  - Interpretation and evaluation requirements
  - Marking, reporting, and other documentation requirements
  - Requirements for instructions including technique sheets and inspection reports
  - Post-inspection requirements

Instruction
Description of the steps to be followed when performing an NDT technique (developed in conformance with a procedure)

Scope

Test sequence

Other categories as necessary

Technique Sheet
- Test part identification
- Authorization
- Inspection areas of interest
- Part diagrams
- Equipment and materials
- Equipment settings
- Details of processing
- Acceptance criteria
- Cleaning/part preparation
- Other unique information
- Author's name, certification, and date

Inspection Report(s)
- Test part identification
- Authorization
- Equipment and materials
- Calibration and verification
- Coverage
- Test results
- Sketch of results
- Other unique information
- Inspector's name, certification, and date

Figure 2.5: PCIP portion of the Level II General Examination in relation to the entire Level II General Examination

Level II Radiography General Examination

Part A (general multiple-choice portion)

- answer 60 multiple-choice questions that assess the candidate's knowledge of fundamentals and principles

Part B (PCIP portion)

Given a procedure and supplementary information:

- complete a technique sheet and an inspection report
- answer 30 multiple-choice questions related to the procedure

Chapter 3
Level II Hands-on Practical Examination

Overview of the Level II Hands-on Practical Examination

The Level II Hands-on Practical Examination assesses the candidate's ability to perform radiographic testing (RT), including the capability to interpret and evaluate test results. There are two categories of Hands-on Practical Examinations, a General Hands-on Practical Examination and a Specific Hands-on Practical Examination. An additional examination associated with the Specific Hands-on Practical Examination is the Specific Written Examination.

General Hands-on Practical Examination

The General Hands-on Practical Examination requires the candidate to set up several radiographic exposures and to interpret and evaluate previously exposed radiographs.

First, the candidate will physically set up typical radiographic exposures in accordance with several different instructions. The instructions will present different exposure arrangements and make use of different equipment and materials. The candidate will not make an actual exposure but will be required to complete a technique sheet documenting significant test parameters for each set up. This part of the practical examination is intended to demonstrate the candidate's ability to perform radiographic inspection over a wide range of applications.

Next, the candidate will interpret and evaluate previously exposed radiographs. Acceptance criteria will be provided for each radiograph (or set of radiographs). The candidate will apply the acceptance criteria and interpret and evaluate the discontinuity indications. This part of the practical examination will confirm the candidate's ability to recognize and evaluate discontinuities associated with the product forms listed in the *Recommended Practice No. SNT-TC-1A* Recommended Training Outline. The key difference between the General Hands-on Practical Examination and the Specific Hands-on Practical Examination is that the General Hands-on Practical Examination is intended to qualify the candidate's ability to perform NDT over a broad range of applications while the Specific Hands-on Practical is specific to an industry sector (i.e., nuclear, aerospace, etc.).

Specific Hands-on Practical Examination

The Specific Hands-on Practical Examination may be required by an industrial sector (an industrial sector is an industry, product form, or area of technology that possesses common characteristics with respect to NDT considerations) and is similar to the General Hands-on Practical Examination except that it is intended to qualify the candidate's ability to perform NDT over a limited range of specialized applications. Specific Hands-on Practical Examinations exist because an industrial sector has unique NDT considerations not normally encountered outside of that particular industry.

For example, if the foundry or metal casting industry determines that NDT technicians perform a limited number of RT techniques to unique codes, standards, or specifications for the purpose of detecting particular discontinuities that have little in common with other product forms or manufacturing processes, then it would be desirable for them to establish an industrial sector. Henceforth, as an industrial sector, there would be a Specific Hands-on Practical Examination that might involve a number of different metal castings (alloy, process, geometry) and different RT techniques.

Specific Written Examination

The Specific Written Examination may be required by an industrial sector in conjunction with either a General Hands-on Practical Examination or Specific Hands-on Practical Examination. The Specific Written Examination assesses a candidate's specific knowledge of practices and techniques unique to the industrial sector, including applicable procedures, codes, standards, specifications, test equipment, and materials.

Continuing with the foundry or metal casting industrial sector example, it would be the Specific Written Examination that would assess a candidate's knowledge and understanding of applicable procedures, codes, standards, specifications, test equipment, and materials as they are used in testing practices and techniques unique to this industrial sector. An industrial sector may even request that this examination be used for assessing the candidate's knowledge of particular variations of industrial sector manufacturing processes as they relate to discontinuities.

Continuing with the foundry or metal casting industrial sector example, casting discontinuities, as a result of green-sand molds, can be quite different than those as a result of dry-sand molds and the candidate may need to know this to accurately perform interpretation and evaluation of casting discontinuity indications.

Appendix 1
Standard Terminology for Gamma and X-radiography

This standard terminology is adapted from ASTM E 1316-94.

Absorbed dose – The amount of energy imparted by ionizing radiation per unit mass of irradiated matter. Denoted by "rad;" 1 rad = 0.01 j/kg. SI unit is "gray;" 1 Gy = 1 j/kg.

Absorbed dose rate – The absorbed dose per unit of time; rads/s. SI unit, grays/s.

Absorption – The process whereby the incident particles or photons of radiation are reduced in number or energy as they pass through matter.

Accelerating potential – The difference in electric potential between the cathode and anode in an X-ray tube through which a charged particle is accelerated: usually expressed in units of kilovoltage or megaelectron volts.

Activation – In neutron radiography, the process of causing a substance to become artificially radioactive by subjecting it to bombardment by neutrons or other particles.

Acute radiation syndrome – The immediate effects of a short term, whole body overexposure of a person to ionizing radiation. These effects include nausea and vomiting, malaise, increased temperature, and blood changes.

Alphanumeric – Term pertaining to both numbers and alphabetical characters typically used to designate a device capable of handling both types of characters.

Alpha particle – A positively-charged particle emitted by certain radio-nuclides. It consists of two protons and two neutrons, and it is identical to the nucleus of a helium atom.

Anode – The positive electrode of a discharge tube. In an X-ray tube, the anode carries the target.

Anode current – The electrons passing from the cathode to the anode in an X-ray tube, minus the small loss incurred by the backscattered fraction.

Aperture – An opening in material, space, or time over which an element is considered to be active.

Array processor – A special purpose logical processing device that performs extremely fast mathematical operations on digital arrays.

Area of interest – The specific portion of the object image on the radiograph that is to be evaluated.

Artifact – Spurious indication on a radiograph arising from, but not limited to, faulty manufacture, storage, handling, exposure, or processing.

Autoradiograph – The image of an object containing a radioelement obtained, on a recording medium, by means of its own radiation.

Betatron – An electron accelerator in which acceleration is provided by a special magnetic field constraining the electrons to a circular orbit. This type of equipment usually operates at energies between 10 and 31 MeV.

Blocking or masking – Surrounding specimens or covering their sections with absorptive material.

Blooming – In radiologic real-time imaging, an undesirable condition exhibited by some image conversion devices and television pickup tubes brought about by exceeding the allowable input brightness for the device, causing the image to go into saturation, producing a fuzzy image of degraded spatial resolution and grayscale rendition.

Blow back – The enlargement of a minified radiograph to its original size by use of an optical direct reader.

Cassette – See **Film holder**.

Characteristic curve – The plot of density versus log of exposure or log of relative exposure. (Also called the D-log E curve or the H & D curve.)

Cine-radiography – The production of a series of radiographs that can be viewed rapidly in sequence, thus creating an illusion of continuity.

Collimator – A device of radiation absorbent material intended for defining the direction and angular divergence of the radiation beam.

Composite viewing – The viewing of two or more superimposed radiographs from a multiple film exposure.

Contrast sensitivity – A measure of the minimum percentage change in an object that produces a perceptible density/brightness change in the radiological image.

Contrast stretch – A function that operates on the grayscale values in an image to increase or decrease image contrast.

Definition, image definition – The sharpness of delineation of image details in a radiograph. Generally used qualitatively.

Densitometer – A device for measuring the optical density of radiographic film.

Density (film) – The quantitative measure of film blackening when light is transmitted or reflected. $D = \log(I_o/I_t)$ or $D = \log(I_o/R)$

where:

D = density,
I_o = light intensity incident on the film,
I_t = light intensity transmitted, and
R = light intensity reflected

Density comparison strip – See **Step-wedge comparison film**.

Digital image acquisition system – A system of electronic components which, by either directly detecting radiation or converting analog radiation detection information, creates an image of the spatial radiation intensity map comprised of an array of discrete digital intensity values (see **Pixel**).

Appendix 1, Standard Terminology for Gamma and X-radiography

Equivalent IQI sensitivity – That thickness of IQI expressed as a percentage of the section thickness radiologically examined in which a 2T hole or 2% wire size equivalent would be visible under the same radiological conditions.

Equivalent penetrameter sensitivity – That thickness of penetrameter, expressed as a percentage of the section thickness radiographed, in which a 2T hole would be visible under the same radiographic conditions.

Erasable optical medium – An erasable and rewritable storage medium where the digital data is represented by the degree of reflectivity of the medium recording layer; the data can be altered.

Exposure, radiographic exposure – The subjection of a recording medium to radiation for the purpose of producing a latent image. Radiographic exposure is commonly expressed in terms of milliampere-seconds or millicurie-hours for a known source-to-film distance.

Exposure table – A summary of values of radiographic exposures suitable for the different thicknesses of a specified material.

Film contrast – A qualitative expression of the slope or steepness of the characteristic curve of a film; that property of a photographic material that is related to the magnitude of the density difference resulting from a given exposure difference.

Film holder – A light-tight container for holding radiographic recording media during exposure, for example, film, with or without intensifying or conversion screens.

Film speed – A numerical value expressing the response of an image receptor to the energy of penetrating radiation under specified conditions.

Filter – Uniform layer of material, usually of higher atomic number than the specimen, placed between the radiation source and the film for the purpose of preferentially absorbing the softer radiations.

Fluorescence – The emission of light by a substance as a result of the absorption of some other radiation of shorter wavelengths only as long as the stimulus producing it is maintained.

Fluorescent screen – Alternative term for intensifying screen (b).

Fluoroscopy – The visual observation on a fluorescent screen of the image of an object exposed to penetrating, ionizing radiation.

Focal spot – For X-ray generators, that area of the anode (target) of an X-ray tube that emits X-rays when bombarded with electrons.

Fog – A general term used to denote any increase in optical density of a processed photographic emulsion caused by anything other than direct action of the image forming radiation and due to one or more of the following:

 a. aging – deterioration, before or after exposure, or both, resulting from a recording medium that has been stored for too long a period of time, or other improper conditions.
 b. base – the minimum uniform density inherent in a processed emulsion without prior exposure.
 c. chemical – resulting from unwanted reactions during chemical processing.
 d. dichroic – characterized by the production of colloidal silver within the developed sensitive layer.

e. exposure – arising from any unwanted exposure of an emulsion to ionizing radiation or light at any time between manufacture and final fixing.
f. oxidation – caused by exposure to air during developing.
g. photographic – arising solely from the properties of an emulsion and the processing conditions, for example, the total effect of inherent fog and chemical fog.
h. threshold – the minimum uniform density inherent in a processed emulsion without prior exposure.

Fog density – A general term used to denote any increase in the optical density of a processed film caused by anything other than the direct action of the image-forming radiation.

Gamma radiography – A technique of producing radiographs using gamma rays.

Gamma ray – Electromagnetic penetrating radiation having its origin in the decay of a radioactive nucleus.

Geometric unsharpness – The penumbral shadow in a radiological image which is dependent upon, 1. the radiation source dimensions, 2. the source-to-object distance, and 3. object-to-detector distance.

Graininess – The visual impression of irregularity of silver deposits in a processed film.

Half-life – The time required for one half of a given number of radioactive atoms to undergo decay.

Half-value layer (HVL) – The thickness of an absorbing material required to reduce the intensity of a beam of incident radiation to one half of its original intensity.

Half-value thickness – The thickness of a specified substance that, when introduced into the path of a given beam of radiation, reduces its intensity to one half.

Image definition – See **Definition**.

Image quality indicator (IQI) – In industrial radiology, a device or combination of devices whose demonstrated image or images provide visual or quantitative data, or both, to determine radiologic quality and sensitivity. Also known as a penetrameter (disparaged).

Note – It is not intended for use in judging size or for establishing acceptance limits of discontinuities.

Indication – The response or evidence from a nondestructive examination that requires interpretation to determine relevance.

Intensifying screen – A material that converts a part of the radiographic energy into light or electrons and that, when in contact with a recording medium during exposure, improves the quality of the radiograph, or reduces the exposure time required to produce a radiograph, or both. Three kinds of screens in common use are:

a. metal screen – a screen consisting of dense metal (usually lead) or of a dense metal compound (for example, lead oxide) that emits primary electrons when exposed to X- or gamma rays.
b. fluorescent screen – a screen consisting of a coating of phosphors that fluoresces when exposed to X- or gamma radiation.
c. fluorescent-metallic screen – a screen consisting of a metallic foil (usually lead) coated with a material that fluoresces when exposed to X- or gamma radiation. The coated surface is placed next to the film to provide fluorescence; the metal functions as a normal metal screen.

Appendix 1, Standard Terminology for Gamma and X-radiography

IQI sensitivity – In radiography, the minimum discernible image and the designated hole in the plaque-type, or the designated wire image in the wire type image quality indicator.

keV (kiloelectron volt) – A unit of energy equal to 1000 electronvolts, used to express the energy of X-rays, gamma rays, electrons, and neutrons.

kV (kilovolt) – A unit of electrical potential difference equal to 1000 V, used to described the accelerating potential of an X-ray tube.

Latent image – A condition produced and persisting in the image receptor by exposure to radiation and able to be converted into a visible image by processing.

Lead screen – see **Intensifying screen (a)**.

Line pair test pattern – A pattern of one or more pairs of objects with high contrast lines of equal width and equal spacing. The pattern is used with an imaging device to measure spatial resolution.

Linear accelerator – An electron generator in which the acceleration of the particles is connected with the propagation of a high-frequency field inside a linear or corrugated waveguide.

Line pairs per millimeter – A measure of the spatial resolution of an image conversion device. A line pair test pattern consisting of one or more pairs of equal width, high contrast lines and spaces is utilized to determine the maximum density of lines and spaces that can be successfully imaged. The value is expressed in line pairs per millimeter.

Location marker – A number or letter made of lead (Pb) or other highly radiation attenuative material that is placed on an object to provide traceability between a specific area on the image and the part.

Low-energy gamma radiation – Gamma radiation having energy less than 200 keV.

Luminosity – A measure of emitted light intensity.

mA (milliampere) – A unit of current equal to 0.001 A, used to express the tube current of an X-ray tube.

Magnetic storage medium – A storage medium that uses magnetic properties (magnetic dipoles) to store digital data (for example, a moving drum disk, tape, static core, or film).

MeV (mega or million electronvolts) – A unit of energy equal to one million electronvolts, used to express the energy of X-rays, gamma rays, electrons, and neutrons.

Microfocus X-ray tube – An X-ray tube having an effective focal spot size not greater than 100 μm.

Milliamperes (mA) – The technical term is tube current and it is defined as the current passing between the cathode and anode during the operation of an X-ray tube, measured in milliamperes (mA) and usually taken as a measure of X-ray intensity.

Minifocus X-ray tube – An X-ray tube having an effective focal spot size between 100 and 400 μm.

MV (mega or million volt) – A unit of electrical potential difference equal to one million volts, used to describe the accelerating potential of an X-ray tube.

Net density – Total density less fog and support (film base) density.

Neutron radiography (NR) – A process of making an image of the internal details of an object by the selective attenuation of a neutron beam by the object.

Noise – The data present in a radiological measurement that is not directly correlated with the degree of radiation attenuation by the object being examined.

Nonerasable optical data – A nonerasable, nonrewritable storage medium where the digital data is represented by the degree of reflectivity of the medium's recording layer. The data cannot be altered.

Nonscreen-type film (direct-type film) – X-ray film designed for use with or without metal screens, but not intended for use with salt screens.

Nuclear activity – The number of disintegrations occurring in a given quantity of material per unit of time. Becquerels (curies) is the unit of measurement.

Object-to-film distance – The distance between the surface of the source side object and the plane of the recording medium.

> **Note** – In the case where the recording medium is placed directly in contact with the object being examined, the distance is equal to the thickness of the object.

Optical density – The degree of opacity of a translucent medium (darkening of film) expressed as follows: $OD = \log(I_o/I_t)$

where:
OD = optical density,
I_o = light intensity incident on the film, and
I_t = light intensity transmitted through the film

Optical line pair test pattern – see **Line pair test pattern**.

Pair production – The process whereby a gamma photon with energy greater than 1 MeV is converted directly into matter in the form of an electron-positron pair. Subsequent annihilation of the positron results in the production of two 0.5 MeV gamma photons.

Penetrameter – See **Image quality indicator**.

Penetrameter sensitivity – See **IQI sensitivity**.

Phosphor – Any substance that can be stimulated to emit light by incident radiation.

Photo fluorography – A photograph of the image formed on a fluorescent screen.

Photostimulable luminescence – The physical phenomenon of phosphors absorbing incident ionizing radiation, storing the energy in quasi-stable states, and emitting luminescent radiation proportional to the absorbed energy when stimulated by radiation of a different wavelength.

Pixel – A short form of picture element. The smallest addressable element in a electronic image.

Pixel, display size – The dimensions of the smallest picture element comprising the displayed image, given in terms of the imaged object's dimensions being represented by the element.

Pixel size – The length and width dimensions of a pixel.

Primary radiation – Radiation coming directly from the source.

Appendix 1, Standard Terminology for Gamma and X-radiography

Radiograph – A permanent, visible image on a recording medium produced by penetrating radiation passing through the material being tested.

Radiographic contrast – The difference in density between an image and its immediate surroundings on a radiograph.

Radiographic equivalence factor – That factor by which the thickness of a material must be multiplied in order to determine what thickness of a standard material (often steel) will have the same absorption.

Radiographic exposure – See **Exposure**.

Radiographic inspection – The use of X-rays or nuclear radiation, or both, to detect discontinuities in material, and to present their images on a recording medium.

Radiographic quality – A qualitative term used to describe the capability of a radiograph to show flaws in the area under examination.

Radiographic sensitivity – A general or qualitative term referring to the size of the smallest detail that can be seen on a radiograph, or the ease with which details can be seen.

Radiological examination – The use of penetrating ionizing radiation to display images for the detection of discontinuities or to help ensure integrity of the part.

Radiology – The science and application of X-rays, gamma rays, neutrons, and other penetrating radiations.

Radioscopy – The electronic production of a radiological image that follows very closely the changes with time of the object being imaged.

Rare earth screens – See **Intensifying screen**.

Real-time radioscopy – Radioscopy that is capable of following the motion of the object without limitation of time.

Recording media – Material capable of capturing and/or storing a radiological image in digital or analog form.

Recording medium – A film or detector that converts radiation into a visible image.

Representative quality indicator (RQI) – An actual part or similar part of comparable geometry and attenuation characteristics to that of the test part(s), that has known and/or measurable features representing the facets of nonconformance for which the test part is to be examined.

Scintillators and scintillating crystals – A detector that converts ionizing radiation to light.

Screen – See **Intensifying screen**.

Secondary radiation – Radiation emitted by any substance as the result of irradiation by the primary source.

Sensitivity – See **Contrast sensitivity, Equivalent IQI sensitivity, Equivalent penetrameter sensitivity, IQI sensitivity,** and **Radiographic sensitivity.**

Shim – A material, typically placed under the IQI, that is radiologically similar to the object being imaged.

Signal – The data present in a radiological measurement that is directly correlated with the degree of radiation attenuation by the object being examined.

Source – A machine or radioactive material that emits penetrating radiation.

Source-to-film distance – The distance between the radiation producing area of the source and the film.

Step wedge – A device with discrete step thickness increments used to obtain an image with discrete density step values.

Step-wedge calibration film – A step-wedge comparison film the densities of which are traceable to a nationally recognized standardizing body.

Step-wedge comparison film – A strip of processed film carrying a stepwise array of increasing photographic density.

Step wedge comparison film – A radiograph with discrete density steps that have been verified by comparison with a calibrated step wedge film.

Subject contrast – The ratio (or the logarithm of the ratio) of the radiation intensities transmitted by selected portions of the specimen.

System induced artifacts – Anomalies that are created by a system during the acquisition, display processing, or storage of a digital image.

System noise – The noise present in a radiological measurement resulting from the individual elements of the radiological system.

Target – That part of the anode of an X-ray emitting tube that is hit by the electron beam.

Tenth-value-layer (TVL) – The thickness of the layer of a specified substance that, when introduced into the path of a given narrow beam of radiation, reduces the intensity of this radiation by a factor of ten.

Tomography – Any radiologic technique that provides an image of a selected plane in an object to the relative exclusion of structures that lie outside the plane of interest.

Total image unsharpness – The blurring of test object features in a radiological image resulting from any cause(s).

Translucent base media – Materials with properties that allow radiological interpretation by transmitted or reflected light.

Transmission densitometer – An instrument that measures the intensity of the transmitted light through a radiographic film and provides a readout of the transmitted film density.

Transmitted film density – The density of radiographic film determined by measuring the transmitted light.

Tube current – The transfer of electricity, created by the flow of electrons, from the filament to the anode target in an X-ray tube; usually expressed in unit of milliamperes.

vacuum cassette – A flexible, light-tight container that, when operated under a vacuum, holds film and screen in intimate contact during a radiographic exposure.

Appendix 2
Qualification and Certification of NDT Personnel

Overview of Personnel Qualification and Certification

Qualification and certification of NDT personnel in the United States (U.S.) has traditionally been through employer-managed programs based on *Recommended Practice No. SNT-TC-1A*, or operated in accordance with the *ASNT Standard for Qualification and Certification of Nondestructive Testing Personnel (ANSI/ASNT CP-189)*. ASNT first issued *SNT-TC-1A* in 1968 while *ANSI/ASNT CP-189* was first issued in 1991. Since 1977, ASNT has actively supported employer-managed NDT qualification and certification programs by offering ASNT NDT Professional Level III certification by examination in various NDT methods to U.S. NDT personnel and to NDT personnel from countries all over the world.

As the global marketplace continues to expand, the need for global standards increases. Global standards help increase harmonization between countries, industries, and technical societies, help facilitate international commerce, and foster mutual acceptance among partners of NDT personnel qualification and certification. With this in mind, ASNT, in 1996, began implementing a new NDT personnel qualification and certification program entitled the ASNT Central Certification Program (ACCP). The purpose of the ACCP is to provide Professional Level III, Level II, and Level I NDT personnel with independent, transportable certification by examination for national and international acceptance. The ACCP is uniquely flexible in that it allows for qualification and certification that satisfies any number of requirements including those of *ISO 9712 - Non-destructive Testing: Qualification and Certification of Personnel*.

The following sections provide an overview of *SNT-TC-1A*, *ANSI/ASNT CP-189*, *ISO 9712* and the ACCP.

Recommended Practice No. SNT-TC-1A

This document is intended to be a guideline for employers to establish their own written practice for the qualification and certification of their NDT personnel. It is not intended to be used as a strict specification. This document was first issued in 1968 and was revised in 1975, 1980, 1984, 1988, 1992, and 1996. The current edition of *SNT-TC-1A*, (1996), includes the following NDT methods:
 1. acoustic emission testing (AE),
 2. electromagnetic testing (ET),
 3. leak testing (LT),
 4. liquid penetrant testing (PT),
 5. magnetic particle testing (MT),
 6. neutron radiographic testing (NR),
 7. radiographic testing (RT),
 8. infrared/thermal testing (IR),
 9. ultrasonic testing (UT),
 10. vibration analysis (VA), and
 11. visual and optical testing (VT).

SNT-TC-1A defines three levels of NDT qualification (Level I, Level II, Professional Level III) as well as the recommended education, training, and experience requirements for each level. It also establishes the different types of examinations for each level of qualification.

1. Professional Level III:
   a. Basic Examination (required only once independent of the number of methods),
   b. Method Examination (for each method), and
   c. Specific Examination (for each method).
2. Level II:
   a. General Examination (for each method),
   b. Specific Examination (for each method), and
   c. Practical Examination (for each method).
3. Level I:
   a. General Examination (for each method),
   b. Specific Examination (for each method), and
   c. Practical Examination (for each method).

SNT-TC-1A recommends the minimum number of questions for each written examination and the format for practical examinations.

The following excerpts from Section 9 of *SNT-TC-1A* present details concerning certification:

9.1 Certification of all levels of NDT personnel is the responsibility of the employer.

9.2 Certification of NDT personnel shall be based on demonstration of satisfactory qualification in accordance with Sections 6, 7, and 8, as modified by the employer's written practice.

9.3 At the option of the employer, an outside agency may be engaged to provide NDT Professional Level III services. In such instances, the responsibility of certification is retained by the employer.

9.4 Personnel certification records shall be maintained on file by the employer and should include the following:
  1. Name of certified individual.
  2. Level of certification and NDT method.
  3. Educational background and experience of certified individuals.
  4. Statement indicating satisfactory completion of training in accordance with the employer's written practice.
  5. Results of the vision examinations prescribed in 8.2 for the current certification period.
  6. Current examination copy(ies) or evidence of successful completion of examinations.
  7. Other suitable evidence of satisfactory qualifications when such qualifications are used in lieu of the specific examination prescribed in 8.8.3(b) or as prescribed in the employer's written practice.
  8. Composite grade(s) or suitable evidence of grades.
  9. Dates of certification and/or recertification and the dates of assignments to NDT.
  10. Signature of employer's certifying authority.

The following portions from Section 9 of *SNT-TC-1A* presents details concerning recertification:

9.5 Recertification
  1. All levels of NDT personnel shall be recertified periodically in accordance with one of the following criteria:
     a. Evidence of continuing satisfactory performance.
     b. Reexamination in those portions of the examinations in Section 8 deemed necessary by the employer's NDT Level III.
  2. Recommended maximum recertification intervals are:
     a. Levels I and II — 3 years, and
     b. Professional Level III — 5 years.
  3. NDT personnel may be reexamined any time at the discretion of the employer and have their certificates extended or revoked.
  4. The employer's written practice should include rules covering the duration of interrupted service that requires reexamination and recertification.

ANSI/ASNT CP-189

This document is the U.S. consensus standard for qualification and certification of NDT personnel. The current edition of *ANSI/ASNT CP-189* (1995) includes the NDT methods of AE, ET, LT, MT, NR, PT, RT, VT and UT. It identifies five categories of NDT qualification (Professional Level III, Level II, Level I, Trainee, NDT Instructor); however, only Professional Level III, Level II and Level I personnel are certified while a qualified NDT Instructor is designated by an NDT Professional Level III.

In much the same manner as *SNT-TC-1A*, *ANSI/ASNT CP-189* requires the employer to establish a "procedure" (*SNT-TC-1A* uses the term "written practice") for the qualification and certification of NDT personnel. This standard defines the education, training, and experience requirements for each category of qualification. It also establishes the different types of examinations for each level of qualification.

1. Professional Level III:
   a. ASNT NDT Professional Level III certificate (with a currently valid endorsement for each method in which employer certification is sought).
   b. Specific Examination (for each method),
   c. Practical Examination (for each method; prepare an NDT procedure), and
   d. Demonstration Examination (for each method; hands-on practical examination).
2. Level II:
   a. General Examination (for each method),
   b. Specific Examination (for each method), and
   c. Practical Examination (for each method).
3. Level I:
   a. General Examination (for each method),
   b. Specific Examination (for each method), and
   c. Practical Examination (for each method).

Candidates who fail an examination are required to receive additional documented training addressing the deficiencies that caused failure, or wait at least thirty days before reexamination. *ANSI/ASNT CP-189* requires an NDT Professional Level III with a valid ASNT NDT Professional Level III certificate in the applicable method to be responsible for development, administration and grading of examinations; however, in no case is it permitted that an examination be prepared or administered by one's self or by one's subordinate.

Similar to *SNT-TC-1A*, this standard specifically provides the employer with the option to engage the services of an outside organization to perform the duties of an NDT Professional Level III. *ANSI/ASNT CP-189* states that the training requirements for an NDT Professional Level III are satisfied if the individual holds a current ASNT NDT Professional Level III certificate in the applicable NDT method.

Section 5 of *ANSI/ASNT CP-189* presents details concerning certification:

5.1 Procedure. The employer shall develop and maintain a procedure detailing the program that will be used for qualification and certification of NDT personnel in accordance with this standard.

5.2 Procedure Requirements. The procedure shall describe the minimum requirements for certifying personnel in each NDT method and the levels of qualification desired. The procedure shall satisfy the requirements of this standard. The procedure shall include, as a minimum, the following:
   a. personnel duties and responsibilities including, if the employer has more than one NDT Professional Level III for a specific method, the employer shall designate one individual as the principal NDT Professional Level III for each such method;
   b. training requirements;
   c. experience requirements;
   d. examination requirements;
   e. records and documentation requirements, including control, responsibility, and retention period; and

f. recertification requirements.

5.3 Approval. The employer's certification procedure shall be approved by an NDT Professional Level III designated by the employer.

Certification also requires successful completion of vision examinations administered in accordance with a procedure, and by personnel approved by the NDT Professional Level III.

Recertification of NDT Level I and Level II personnel is required to be by examination in accordance with certification examination requirements, at least every three years. Recertification of NDT Professional Level III personnel, as a minimum, requires verification of the individual's ASNT NDT Professional Level III certificate for currency in each method for which recertification is sought.

ISO 9712

ISO 9712 is an international standard that establishes a system for the qualification and certification of personnel to perform industrial NDT. Instead of employer-managed certification programs, *ISO 9712* requires certification be conducted by a central, independent body that must be a nonprofit organization with no direct involvement in the training of NDT personnel and that is recognized by the NDT community or the ISO member body of the country concerned.

ISO 9712 introduces a certification process that uses a national certifying body to administer procedures for certification of NDT personnel, and a qualifying body authorized by the national certifying body, to prepare and administer certification examinations. An examination center may be authorized by the national certifying body, or through a qualifying body to administer certification examinations. *ISO 9712* uses the term "industrial sector" to describe an area of industry or technology using specialized NDT that requires specific skill, knowledge, equipment, or training to achieve satisfactory performance.

ISO 9712 defines three levels of NDT qualification (Level 1, Level 2, Level 3) as well as the education, training, and experience requirements for each level of qualification. It also establishes the different types of examinations for each level of qualification.

1. Level 3:
    a. Basic Examination (required only once independent of the number of methods),
    b. Method Examination [for each method; integrating application of the method to the applicable industrial sector(s), and includes drafting one or more procedures in the applicable industrial sector(s)], and
    c. Practical Examination (for each method; Level 2 Hands-on Practical Examination is required when the Level 3 candidate does not hold appropriate Level 2 certification).
2. Level 2:
    a. General Examination (for each method),
    b. Specific Examination [for each method; related to the applicable industrial sector(s)], and
    c. Practical Examination [for each method; related to the applicable industrial sector(s)].
3. Level 1:
    a. General Examination (for each method),
    b. Specific Examination [for each method; related to the applicable industrial sector(s)], and
    c. Practical Examination [for each method; related to the applicable industrial sector(s)].

Certification requires documented evidence of satisfactory vision in accordance with requirements listed in *ISO 9712*. There are a number of requirements associated with validity of certification including no significant interruption of work in the method(s) for which one is certified.

Recertification requirements include continued satisfactory work activity relevant to certification without significant interruption. Recertification is required at least every five years from the date of certification. Every other recertification period, or at least every ten years, the certified individual is also required to pass a limited practical examination if Level I or II, or a written examination if Professional Level III.

ASNT Central Certification Program (ACCP)

The ACCP was adopted by the ASNT Board of Directors 13 July 1996. The ACCP establishes the system for central certification of NDT personnel administered and maintained by ASNT. The purpose of ACCP is to provide independent, transportable NDT certification by examination to promote national and international acceptance of NDT certification and reduce multiple audits of certification programs. The ACCP was developed to improve NDT reliability and accuracy through enhanced performance of personnel as demonstrated by the ACCP examinations and accompanying qualification requirements. The ACCP is intended to provide customers and prospective employers with clear expectations of NDT personnel competency and proficiency.

Management of the ACCP is the responsibility of the Certification Management Council (CMC), which is a standing committee of ASNT. An authorized qualifying body (AQB) may be used to prepare and administer NDT qualification examinations and an authorized examination center (AEC) may be used to administer NDT qualification examinations.

Within the ACCP there are several options available to satisfy employer, employee and industry needs. The ACCP is unique in that it allows for qualification and certification that addresses any number of requirements including those of *SNT-TC-1A, ANSI/ASNT CP-189, MIL-STD-410, ISO 9712*, and other international sources of NDT qualification and certification programs based on *ISO 9712*. It also provides a mechanism for specific practical and written examinations to better accommodate industries where product-forms or areas of technology demand specialized NDT. The ACCP uses a combination of traditional practices, national and international conventions and newly evolved concepts to create a system of central certification which represents the next generation of NDT personnel qualification and certification.

ACCP terminology is of key importance; the following are definitions from the ACCP document:

2.1 *ACCP certification*: The process whereby ASNT certifies that an individual has met the requirements of this document for ACCP Professional Level III, ACCP Level II, or ACCP Level I.

2.2 *ASNT NDT Professional Level III*: An individual who, having passed ASNT administered Basic and Method(s) Examinations, holds a current, valid ASNT NDT Professional Level III certificate in at least one method.

2.3 *Authorized examination center* (AEC): A site with facilities and personnel, independent of the employer, approved by the ASNT Certification Management Council (CMC) to administer NDT qualification examinations.

2.4 *Authorized qualifying body (AQB)*: A competent organization, independent of the employer, approved by the ASNT CMC to prepare and administer NDT qualification examinations.

2.5 *Certificate*: Written testimony of qualification.

2.6 *Certification Management Council (CMC)*: Formerly known as the Certification Management Board (CMB), this is a council of ASNT that is responsible for managing the ACCP.

2.7 *Employer*: The corporate, private, or public entity that employs personnel for wages or salary.

2.8 *Employer authorization*: The process whereby an employer's ACCP Professional Level III or ASNT NDT Professional Level III reviews the certificates of ASNT Central Certification for the employer's NDT personnel, determines if further examination (see job specific examinations in paragraph 7.5.) is required, and then, on behalf of the employer, authorizes personnel to perform NDT for that employer.

2.9 *Endorsement*: Written testimony of a particular qualification.

2.10 *Guidance*: See *supervision*.

2.11 *Industrial sector (IS)*: An industry, product-form or area of technology that possesses common characteristics with respect to NDT considerations.

2.12 *Instruction*: A description of the steps to be followed when performing an NDT technique; developed in conformance with a procedure.

2.13 *Procedure*: A written description that establishes minimum requirements for performing an NDT method on any object, written in accordance with established standards, codes, or specifications.
2.14 *Qualification*: Demonstration or possession of education, skills, training, knowledge, and experience required for personnel to properly perform NDT to a level as specified in this document.
2.15 *Recertification examination*: An examination administered by the CMC expressly for the purpose of recertification.
2.16 *Recertification*: The process of extending one's certification after the initial period of validity, and maintaining certification for individual periods thereafter.
2.17 *Renewal*: Same as recertification.
2.18 *Specific procedure*: Same as instruction.
2.19 *Supervision*: The act of an ACCP Level II, ACCP Professional Level III, or ASNT NDT Professional Level III directing the application of NDT performed by other NDT personnel, which includes the control of actions involved in the preparation of the test, performance of the test, and reporting of test results.

The ACCP refers to a number of appendices, each traceable to a source (e.g., *SNT-TC-1A, ANSI/ASNT CP-189, ISO 9712*, etc.) of differing qualification requirements (education, training, and experience). There are three levels of qualification: Professional Level III, Level II, and Level I. The examinations for each level of qualification are:
1. Professional Level III: candidates with a current, valid ASNT NDT Professional Level III certificate in a method shall be considered to have met all prerequisites, except vision, in that method and to have passed the Basic Examination and the Method Examination.
   a. Basic Examination (required only once independent of the number of methods).
   b. Method Examination (for each method).
   c. Procedure Preparation Examination (for each method).
   d. Hands-on Practical Examination (for each method; general or specific as applicable).
2. Level II:
   a. General Examination (for each method; includes Procedure Comprehension and Instruction Preparation (PCIP) portion).
   b. Hands-on Practical Examination (for each method; general or specific as applicable).
3. Level I:
   a. General Examination (for each method; includes Instruction Comprehension portion).
   b. Hands-on Practical Examination (for each method; general or specific as applicable).

Certification requires documented evidence of satisfactory vision in accordance with requirements listed in the ACCP document. There are a number of requirements associated with validity of certification including no significant interruption of work in the method(s) for which one is certified.

Recertification requirements include continued satisfactory work activity, relevant to certification, without significant interruption. Recertification is required at least every five years from the date of certification. Every other recertification period, or at least every ten years, the certified individual is also required to pass a recertification examination applicable to the level of recertification.